MIDLAN[D]
• TIMES •

CONTENTS

Introduction	3
LMS Constituent Companies	4-7
Duchesses go a wandering	8-13
Claughton accident at Culgaith	14-19
London Electrification Schemes	20-25
LMR 4-8-4 Fell Locomotive No. 10100	26-37
The Stanmore Branch	38-41
Operational problems of the LMS, 1939-1945	42-51
Stockport Edgeley	52
The Charnwood Forester	53
Turning the tables at Green Park	54-58
LMS Gas Storage Tanks	59
The one and only 'Big Bertha'	60-67
Bangor station	68-71
West End of Edinburgh	72-79
The Platform End	80

ISBN: 978-1-913251-42-0
First published in 2022 by Transport Treasury Publishing Ltd., 16 Highworth Close, High Wycombe HP13 7PJ

Copies of many of the images in MIDLAND TIMES are available for purchase/download.
In addition the Transport Treasury Archive contains tens of thousands of other UK, Irish and some European railway photographs.

www.ttpublishing.co.uk or for editorial issues and contributions email **MidlandTimes1884@gmail.com**

Printed in Malta by Gutenberg Press.

INTRODUCTION

Welcome to the first issue of Midland Times from Transport Treasury Publishing and we hope that you enjoy the content in this periodical.

I'll just give you a brief bit of information about the editor. My name is Peter Sikes, my late father was a signalman in the Leicester area for almost 40 years, he lived and breathed railways and my interest in the subject obviously stems from him. He joined the LMS in 1947 after army service and worked on the London Midland Region until retirement in 1985. I am also a volunteer for The LMS-Patriot Project, producing our quarterly magazine – The Warrior – and I also work on the fund-raising sales stand when time permits.

The remit for the magazine was to write articles on the London, Midland & Scottish Railway, its constituent companies and the London Midland Region of British Railways, and over the ensuing issues of this periodical we will endeavour to cover as much as we can. There's definitely plenty to research judging by the list of constituent and subsidiary companies that are listed on the following four pages and I'd like to thank those who have contributed to this issue.

So, to start off in this issue we have two in-depth articles featuring one-off locomotives the 'Lickey Banker' and the Fell diesel, both of which were unique in their own way. Added to those we cover an early electrification scheme and how the LMS fared in the dark days of World War II – variety is what we are aiming for.

The articles are accompanied by some fabulous photography that is held in The Transport Treasury archives and most of these are available for you to purchase in differing formats. Contact details can be found on the opposite page.

I hope you will enjoy what you see and read and would encourage you to point out anything that may be erroneous, and indeed, if you feel as though you have some interesting material to share with us don't hesitate to send in your contributions.

PETER SIKES, EDITOR, MIDLAND TIMES
email: midlandtimes1884@gmail.com

FRONT COVER (AND INSET RIGHT): Princess Coronation Class 4-6-2 No. 6221 QUEEN ELIZABETH pictured in all its glory and gaining admiration from onlookers at Polmadie shed, Glasgow in 1937.

ABOVE LEFT: Fresh from a mechanical overhaul Stanier Jubilee No. 45671 PRINCE RUPERT, in an unusual pink undercoat, stands outside the paint shop at Crewe on 19th March 1960. To the left is Fairburn Class 4MT 2-6-4T No. 42674 looking pristine after its repaint.

LMS CONSTITUENT COMPANIES

The following made up the London, Midland and Scottish Railway as a result of the Railways Act 1921:

CONSTITUENT COMPANIES		
Railway Company	Route miles/km	Notes
Caledonian Railway	1,114½ miles (1,794km)	
Furness Railway	158 miles (254km)	
Glasgow and South Western Railway	493½ miles (794km)	
Highland Railway	506 miles (814km)	
London and North Western Railway	2,667½ miles (4,293km)	Amalgamated with the Lancashire and Yorkshire Railway (L&YR) on 1st January 1922
Midland Railway	2,170¾ miles (3,493km)	
North Staffordshire Railway	220¾ miles (355km)	

SUBSIDIARY COMPANIES Independently operated lines		
Railway Company	Route miles/km	Notes
Arbroath and Forfar Railway	14¾ miles (24km)	Leased to or worked by Caledonian Railway
Brechin and Edzell District Railway	6¼ miles (10km)	Leased to or worked by Caledonian Railway
Callander and Oban Railway	99¾ miles (161km)	Leased to or worked by Caledonian Railway
Cathcart District Railway	5¼ miles (8.5km)	
Charnwood Forest Railway	10½ miles (17km)	Leased to or worked by L&NWR
Cleator and Workington Junction Railway	30½ miles (49km)	Partially worked by the Furness Railway
Cockermouth, Keswick and Penrith Railway	30¾ miles (49.5km)	Independent line with rolling stock provided by other companies
Dearne Valley Railway*	21 miles (34km)	Leased to or worked by L&NWR
Dornoch Light Railway	7¾ miles (12km)	Leased to or worked by Highland Railway
Dundee and Newtyle Railway	14½ miles (23km)	Leased to or worked by Caledonian Railway
Harborne Railway	2½ miles (4km)	Leased to or worked by L&NWR
Killin Railway	5¼ miles (8km)	Leased to or worked by Caledonian Railway
Knott End Railway	11½ miles (19km)	Independently operated line
Lanarkshire and Ayrshire Railway	36¼ miles (58km)	Leased to or worked by Caledonian Railway
Leek and Manifold Valley Light Railway	8¼ miles (13km)	Narrow gauge. Originally leased to or worked by North Staffordshire Railway
Maryport and Carlisle Railway	42¾ miles (68.8km)	Independently operated line
Mold and Denbigh Junction Railway	15 miles (24km)	Leased to or worked by L&NWR
North & South Western Junction Railway	5¼ miles (8km)	Worked by several component companies
North London Railway*	16 miles (26km)	Managed by the L&NWR
Portpatrick and Wigtownshire Joint Railway	82¼ miles (132km)	Worked by several component companies
Shropshire Union Railways and Canal Company	29¼ miles (47km)	Originally leased to or worked by L&NWR. Part of this system was jointly leased with the Great Western Railway
Solway Junction Railway	12¼ miles (20km)	Leased to or worked by Caledonian Railway
Stratford-upon-Avon and Midland Junction Railway	67½ miles (109km)	Independently operated line
Tottenham and Forest Gate Railway	6 miles (10km)	Leased to or worked by Midland Railway
Wick and Lybster Light Railway	13½ miles (21.75km)	Leased to or worked by Highland Railway
Wirral Railway	13¾ miles (22km)	Independently operated line
Yorkshire Dales Railway	9 miles (14km)	Leased to or worked by Midland Railway

*Incorporated with the L&NWR Company as from 1st January 1922.

LEFT: Examples of LMS constituent company locomotives can still be seen working today. Here we see Caledonian Railway Class 439 0-4-4T No. 419 at Shackerstone while on a visit to The Battlefield Line from the Scottish Railway Preservation Society based in Bo'ness, on 2nd July 2022. Designed by J. F. McIntosh and built at St. Rollox Works, Glasgow in 1907. The class was designed for branch line work, fast suburban and banking duties. PHOTO: © PETER SIKES

JOINT RAILWAYS After 1923 amalgamations comprised wholly in the LMS		
Railway Company	Route miles/km	Notes
Ashby and Nuneaton Railway	29¼ miles (47km)	L&NWR/Midland Railway joint
Carlisle Citadel Station and Goods Traffic Joint Committees		Jointly owned by various companies
Enderby Railway	2¼ miles (4km)	L&NWR/Midland Railway joint
Furness and Midland Joint Railway	9¾ miles (16km)	Furness/Midland Railway joint
Glasgow, Barrhead and Kilmarnock Joint Railway	29¾ miles (48km)	Caledonian Railway/G&SWR joint
Glasgow and Paisley Joint Railway	14¼ miles (23km)	Caledonian Railway/G&SWR joint
Lancashire Union Railway	12¾ miles (21km)	L&NWR/L&YR joint
North Union Railway	6½ miles (10km)	L&NWR/L&YR joint
Preston and Longridge Railway	8 miles (13km)	L&NWR/L&YR joint
Preston and Wyre Joint Railway	46 miles (74km)	L&NWR/L&YR joint
Whitehaven Cleator and Egremont Railway	35 miles (56km)	Furness Railway/L&NWR joint
JOINT RAILWAYS After 1923 amalgamations joint with the London & North Eastern Railway		
Axholme Joint Railway	27¾ miles (45km)	
Cheshire Lines Committee	142 miles (229km)	One-third share
City of Glasgow Union Railway		
Dumbarton & Balloch	7 miles (11km)	Including Loch Lomond steamers
Dundee and Arbroath Railway	23 miles (37km)	Including Carmyllie Light Railway
Great Central and Midland Joint Railway	40¼ miles (65km)	
Great Central, Hull & Barnsley and Midland Joint Railway	4 miles (6km)	One-third share
Great Central & North Staffordshire Joint Railway	11 miles (18km)	
Great Northern and London & North Western Joint Railway	45 miles (72km)	
Halifax and Ovenden Railway	2½ miles (4km)	
Halifax High Level	3 miles (5km)	
Manchester South Junction and Altrincham Railway	9½ miles (15km)	Joint working arrangement
Methley Joint Line	6 miles (10km)	
Midland and Great Northern Joint Railway	183¼ miles (295km)	
Norfolk and Suffolk Joint Railway	22½ miles (36km)	Great Eastern Railway/Midland Railway/ Great Northern Railway joint
Oldham, Ashton and Guide Bridge Railway	6¼ miles (10km)	
Otley and Ilkley Joint Railway	6¼ miles (10km)	
Perth General Station Committee	–	Two-thirds share

JOINT RAILWAYS After 1923 amalgamations joint with the London & North Eastern Railway (continued)		
Railway Company	Route miles/km	Notes
Prince's Dock, Glasgow	1¼ miles (2km)	
South Yorkshire Joint Railway	20½ miles (33km)	Two-fifths share
Swinton and Knottingley Joint Railway	19½ miles (31km)	
Tottenham and Hampstead Junction Railway	4¾ miles (8km)	
JOINT RAILWAYS After 1923 amalgamations joint with the Great Western Railway		
Birkenhead Railway	56½ miles (91km)	
Brecon & Merthyr Railway and London & North Western Joint Railway	6 miles (10km)	
Brynmawr and Western Valleys Railway	1¼ miles (2km)	
Clee Hill Railway	6 miles (10km)	
Clifton Extension Railway	9 miles (14km)	
Halesowen Joint Railway	6 miles (10km)	
Nantybwch and Rhymney Railway	3 miles (5km)	
Severn and Wye Railway	39 miles (63km)	
Shrewsbury and Hereford Railway	82¾ miles (133km)	
Tenbury Railway	5 miles (8km)	
Vale of Towy Railway	11 miles (18km)	Owned by GWR but leased jointly
West London Railway	2¼ miles (4km)	
Wrexham and Minera Railway	3 miles (5km)	
JOINT RAILWAYS After 1923 amalgamations joint with the Southern Railway		
Somerset and Dorset Joint Railway	105 miles (169km)	
JOINT RAILWAYS After 1923 amalgamations joint with the Metropolitan District Railway		
Whitechapel & Bow Railway	2 miles (3km)	
IRISH RAILWAYS The Railways Act 1921 did not extend to Ireland, but Irish lines owned by constituent companies became part of the LMS		
County Donegal Railways Joint Committee lines	91 miles (146km)	Operated jointly by the NCC and Great Northern Railway of Ireland (GNR(I)), these became joint lines of the LMS and GNR(I) after grouping
Dundalk, Newry and Greenore Railway	26½ miles (43km)	Owned by the LNWR – operated from 1933 by the GNR(I)
Northern Counties Committee lines (NCC)	280¾ miles** (452km)	Owned by the Midland Railway **202½ miles – 5ft 3in gauge 77¼ miles – 3ft gauge
Great Northern Railway (Ireland)	–	Minority shareholding
Great Southern Railways	–	Minority shareholding

DUCHESSES GO A WANDERING

Of the four principal express classes operated by the pre-nationalisation companies – 'Duchess', 'A4', 'King', and 'Merchant Navy' – one type stands out as having visited more areas of the UK (and abroad) than any other, and that was the LMS 'Duchess' (for 'Duchess' of course also read 'Princess Coronation') type. LNER fans please note, the 'A4' comes a very close second.

Starting in 1939, No. 6229 *Duchess of Hamilton* – masquerading as No. 6220 *Coronation* – was shipped from Southampton for a tour of the USA. The engine returned to the UK and original identities were correctly restored.

Next in 1948 came the locomotive exchanges in which No. 46236 *City of Bradford* performed admirably, and uniquely also on all four regions, certainly the only one of the principal express types to do this on so many different metals.

A member of the class, well actually more than one, was back on the Western Region in 1955 and 1956 when Nos. 46237 *City of Bristol* and 46257 *City of Salford* respectively were temporarily loaned to the WR (along with a 'Princess' and various Standard Class 5s from Nine Elms) to cover for 'Kings' temporarily withdrawn due to fatigue cracks found in the latter's front bogies. It is interesting to consider why an ex-LNER type was not similarly loaned, or for that matter a Bulleid design?

Move forward into the 1960s and the 38 members of the class had remained intact up to the end of 1962, although ironically in the same month when the last of the GWR 'Kings' were withdrawn, Nos. 46227 *Duchess of Devonshire*, 46231 *Duchess of Atholl* and 46232 *Duchess of Montrose* were taken out of service for no other reason than necessary repairs were not authorised. Under the same criteria further withdrawals continued through 1963 so that by the start of 1964 half the class had been laid aside with the operators now struggling to find work for an express passenger type whose work had been taken over by less powerful Type 4 2,000hp diesels, a.k.a. the Class 40. Meanwhile aside from filling in for a failed diesel, the class worked parcels, empty stock, specials and even at times fitted freight.

Two options were considered. The first was to replace the last A4s which had themselves found work between Edinburgh and Aberdeen. This was vetoed for nothing more than crew familiarity which itself sounds more than strange when in steam days little in the way of formal training was given to a crew if presented with an unfamiliar engine. Perhaps a more likely scenario was the ER were unwilling to take on additional steam engines when at the same time they were attempting to diverse themselves of as many steam types as possible and similarly as quickly as possible.

The second alternative was on the Southern Region working between Waterloo and Bournemouth. Locally on the SR, inroads had already been made into the numbers of available 'Merchant Navy' class engines and as elsewhere for the simple reason of repairs not being authorised. There was then a shortage of 'Class 8' motive power working out of Waterloo. The 'Duchess' would certainly have fitted the bill with the idea that they would be attached to WD type 8-wheel tenders to increase their water capacity – exactly as had been done to No. 46236 *City of Bradford* when working between Waterloo and Exeter in 1948. (Increased water capacity was needed owing to the lack of water troughs on the Southern.) Steam loco working between Waterloo and Exeter (Central) was soon to be in the hands of the WR so Exeter trains were not an option.

Unfortunately, the Bournemouth line presented two specific gauging issues, the flyover at Battledown west of Basingstoke, the point of divergence of the Bournemouth and West of England lines, and also clearance issues in Southampton tunnel. Both of these would be enough to prevent any transfer although it should be mentioned it is not believed a physical gauging trial with a class member was ever carried out.

No. 6220 CORONATION (alias No. 6229) had been shipped from Southampton to the USA in 1939. This meant it would have traversed the Down Southern main line to reach the port but what we do not know is if, for an engine of its size, it was free from restrictions between Basingstoke and Southampton. (The Southern line between Waterloo and Basingstoke and on to Exeter had not been a problem for the class years later during the 1948 interchange trials.) Rumour has it that any clearance issues would have occurred in Southampton tunnel and also in the Up direction on the flyover at Battledown; neither an issue to No. 6220 in 1939. PHOTO: BILL WHITE

Meanwhile on 9th May 1964 another member of the class returned to the Western Region as part of a railtour. This was the famed RCTS 'East Midlander', No. 46251 *City of Nottingham* taking the train from its starting point at Nottingham Victoria, then south along the Great Central and on to Banbury and Didcot. The tour then continued south to the Southern Region behind No 34038 *Lynton* and eventually back to Swindon where No 46251 was waiting for a similar return to Nottingham.

Sad to report 18 engines still with useful work ahead were withdrawn en masse on 12th September 1964. The survivor was No. 46256 *Sir William A Stanier F.R.S.* retained for one last special working on 26th September between Crewe and Carlisle and vice versa, a tour which also included Nos. 60007 *Sir Nigel Gresley* and 60009 *Union of South Africa* on the Waverley route. Exactly one week later on 3rd October 1964 No. 46256 succumbed. With its historic connections this was an engine that really should have been saved but it was not to be and No. 46256 had been broken up by Cashmore's at their Great Bridge site before the end of 1964.

- Digressing slightly, in 1953 when No. 35020 broke a crank axle at Crewkerne, Eastern Region V2 types were borrowed and apparently well liked by the Southern. For reasons that are again unclear, no reference has been found to redundant engines of this (or any other ER) type being considered for transfer around this time. Perhaps condition was again the issue.

9
470

ABOVE: No. 46236 CITY OF BRADFORD in de-streamlined condition (note the slope on the top of the smokebox), leaving Kings Cross on a familiarisation run 20th April 1948. The two formal tests over the ER with the NER dynamometer car took place on 6th and 7th May 1948. The first day was King's Cross to Leeds with the return the following day. Later that month it undertook similar testing on the GW main line and in June was active between Waterloo and Exeter. The LMR tests had been in April working the 'Royal Scot'. PHOTO: R. E. VINCENT/TRANSPORT TREASURY

LEFT: No. 46257 CITY OF SALFORD after arrival at Paddington with '470', on 21st April 1955, the 4.15pm (SX) Bristol Temple Meads to Paddington. In general terms, it was really only the standard types of steam designs that were universally seen across the regions. Odd transfers, some temporary, some permanent did take place; movement of V2s to the SR has already been mentioned, whilst WR pannier tanks worked at Folkestone, Waterloo and even in Scotland, and Bulleid Pacifics were also on the ER for a short time. There were other examples, especially during WW2.
PHOTO: R. C. RILEY/TRANSPORT TREASURY

In sparkling condition, No. 46251 CITY OF NOTTINGHAM stands at Swindon on 9th May 1964 awaiting the return of the 'East Midlander' tour party from the Southern Region. The engine is in excellent external condition; far better than the 'modern' diesel shunters behind. The same day had been busy for railtours across the network with the abortive Ian Allan high speed run to Plymouth behind No. 4079 PENDENNIS CASTLE, 'Peglar's Pullman' with No. 4472 FLYING SCOTSMAN and

No. 60009 UNION OF SOUTH AFRICA, and finally a rather more down to earth pair of brake vans running around the Bolton area behind a 3F, No. 47378. The same engine, along with No. 46255 CITY OF HEREFORD, was active on another tour on 12th July 1964. It would survive, on paper at least, until 12th September but like many other steam types may well have

CLAUGHTON ACCIDENT AT CULGAITH

6TH MARCH 1930

The accident at Culgaith involved a local passenger train headed by Claughton Class 4-6-0 No. 5971 *Croxteth* leaving Culgaith station, passing the down home signal at danger, and colliding with a stationary ballast train in Waste Bank tunnel which is about three quarters of a mile from the station. As ever, the accident involved a series of unfortunate coincidences which resulted in the crash.

The signalmen at Culgaith, (including those that were off duty) were called to the inquiry as witnesses. The duty signalman reported that *Croxteth*, which was being driven by the fireman, actually hit the tunnel roof, leaving a noticeable mark which may still be there. Unfortunately the driver was killed and his body was left overnight in the ladies waiting room at Culgaith station. The ballast train engine was a Midland 4F which survived the crash and was photographed many years later stuck in a snow drift on the Keswick line.

The damaged loco remained sheeted over in the siding at Culgaith for some time afterwards. There was a detailed article in the Cumberland and Westmorland Herald local newspaper reporting on the inquiry into the accident which was held in Culgaith village hall.

The first three pages (of eight) of The Ministry of Transport crash report are reproduced on the following pages, these outline the details of the incident.

Culgaith station pictured on an unknown date but thought to be some time before 1908. The crossing gates are across the road and a 'down' train is signalled. What was to become Mrs. Miller's restaurant is sporting a tall chimney.

LONDON, MIDLAND AND SCOTTISH RAILWAY.

Ministry of Transport,
4, Whitehall Gardens,
London, S.W.1.
24th April, 1930.

SIR,

I have the honour to report for the information of the Minister of Transport, in accordance with the Order of the 8th March, the result of my Inquiry into the collision which took place at about 10.10 a.m. on the 6th March, between Culgaith and Langwathby, on the Settle-Carlisle section of the London, Midland and Scottish Railway.

The 8.5 a.m. down passenger train, Hellifield to Carlisle, came into head-on collision with a ballast train which was stationary on the down track in Waste Bank Tunnel. I regret to state that the driver of the passenger train was killed, and one passenger died in hospital subsequently as the result of injuries. The fireman, one passenger, and two of the Company's servants travelling as passengers, received serious injuries, and one passenger and one Company's servant travelling as a passenger suffered from minor injuries or shock. In addition to these the guard of the passenger train and the flagman of the ballast train suffered from minor injuries.

The passenger train, which contained about 25 passengers, consisted of four 8-wheeled bogie coaches and one 6-wheeled brake van, the latter being the second vehicle on the train. All vehicles were fitted with the vacuum brake ; the leading and rear coaches were electrically lighted, the three others being gas lighted. The weight of the train was approximately 108 tons. It was drawn by engine No. 5971, type 4–6–0, Claughton class, with 6-wheeled tender, weighing 128 tons in working order, and fitted with the vacuum brake on driving and tender wheels. The overall length of engine and train was 316 ft.

The ballast train consisted of 10 empty ballast wagons, three wagons loaded with new ballast, one 10-ton ballast brake, one 20-ton brake, one tool wagon and one 10-ton ballast brake, in the above order from the engine.

The engine was No. 4009, type 0–6–0, with 6-wheeled tender, weighing 90 tons in working order and fitted with the steam brake, controlled by the vacuum, on all wheels of engine and tender. The overall length of the train was about 373 feet.

As a result of the collision the passenger train engine was turned on its side towards the up line and considerably damaged, the bogie being torn off. The main frames were badly buckled and the boiler moved forward in the frames for about one inch. The engine and tender of the goods train were driven back some 20 yards and also turned over on to the same side and damaged, though to a much less extent than the passenger engine, the main frames being buckled at the front and the smokebox front driven in. The leading coach of the passenger train was slightly damaged, the leading bogie being displaced and driven forward beyond the head-stock, the second coach, a 6-wheeled brake van, was completely destroyed, and the debris of this telescoped and wrecked the leading half of the third coach body. The underframe and leading bogie of this coach were damaged and the trailing half of the body was driven backwards about 18 ins. on the underframe. The two coaches in rear were undamaged.

Of the ballast train three ballast wagons were completely destroyed and four were damaged. Considerable damage was also done to the permanent way. Both up and down lines were blocked and owing to the difficulty of clearing the wreckage in the tunnel they remained blocked for about 48 hours.

A heavy rainstorm came on shortly before the accident. There was little wind ; visibility was good.

Description.

The line in question is the old Midland main line to Scotland via Hellifield and Carlisle and the direction from Culgaith is generally northerly to Langwathby, running along the western slope of the hills.

7120 A

Culgaith signal box is on the up (eastern) side of the line immediately south of a level crossing at the south end of the station.

From this point the line is straight for about 500 yards to the south end of Culgaith tunnel and continues straight through the tunnel for 660 yards; it then takes a slight left-hand curve for 440 yards to the south end of Waste Bank tunnel. The curvature reverses slightly to the right hand through this tunnel which is 164 yards long.

Distances from Culgaith signal box :—

Down home signal	186	yards south
North end of down platform	173	,, north
Down starting signal	304	,, ,,
South end of Culgaith tunnel	496	,, ,,
North end of Culgaith tunnel	1,157	,, ,,
Culgaith up distant signal	1,290	,, ,,
South end of Waste Bank tunnel	1,594	,, ,,
Probable point of collision	1,664	,, ,,
North end of Waste Bank tunnel	1,758	,, ,,
Langwathby signal box	3½	miles ,,

The line falls at 1 in 174 and 1 in 338 through Culgaith station to approximately the starting signal; thereafter gradients are negligible.

The Culgaith down home and starting signals are conspicuously visible on the down side of the line without any obstruction to view. Culgaith up distant signal is on the up side of the line and is provided with a backboard.

Report.

The ballast train from Carlisle arrived at Culgaith at 8.47 a.m. for work in connection with renewal of track in Waste Bank tunnel, the actual work being loading up of spent ballast from the down track. The train was crossed over to the down track at Culgaith and propelled thence to Waste Bank tunnel, signalman Wileman at Culgaith having offered to, and obtained acceptance from, Langwathby, for a ballast train working in section, and the block instruments (of the ordinary three-position type) being put to " train on line."

Before leaving Culgaith, flagman Taylor went to the box, and after informing the signalman of the work to be done, enquired about the probable time available for occupation of the section, the intention being that when it was necessary to clear the line for a running train, the ballast train would propel on to Langwathby, shunt there, work back to Culgaith on the up road, and again propel along the down road to the site of the work.

The next down passenger train was the 8.5 a.m. Hellifield to Carlisle which was due to stop at Culgaith at 10.4 a.m. but the 5.10 a.m. mineral train, Skipton to Carlisle, might or might not arrive ahead of the passenger. According to schedule timings the mineral train should run ahead of the passenger all the way to Carlisle, but in actual working it appears that the passenger train frequently passes the mineral train en route so that it was uncertain whether the ballast train would have to make way for the mineral train, or, alternatively, could have occupation until the passenger train approached. Signalman Wileman made enquiry by telephone from Appleby North Junction box and was informed that there was no goods train north of Ormside (some 9 miles south).

Flagman Taylor then arranged with signalman Wileman that the latter would " shake " the up distant signal as a warning to the ballast train to move on, if the down line was required for a running train. This signal is clearly visible from the greater portion of Waste Bank tunnel. There is a discrepancy of evidence between signalman Wileman, flagman Taylor, and goods guard Proudfoot in this connection. The last named did not come into the box, but was informed of the arrangements by Taylor before leaving Culgaith. Proudfoot understood that the up distant signal would be " shaken " on the approach of any down train, whereas Wileman and Taylor intended the arrangement to hold only in case the mineral train arrived ahead of the passenger. I discuss this arrangement and misunderstanding below.

After informing guard Proudfoot of the arrangement Taylor rode on the engine as far as Waste Bank tunnel and told driver Leach where to stop for the work in hand. He then alighted and went to protect the train, proceeding for three-quarters of a mile along the track towards Culgaith, putting down three detonators at the north end of Culgaith tunnel, and three detonators at the ¾ mile distance, which was between the south end of Culgaith tunnel and the Culgaith down starting signal ; he then remained at this point exhibiting a red flag. It may be noted that this point is in full view from Culgaith station and signal box. Flagman Taylor's actions in this respect were strictly in accordance with regulations.

As the train was stationary in a tunnel, Rule 251 (*g*) (see Appendix) directs the guard to send out a second flagman to put down two detonators and exhibit a red flag at a quarter of a mile in rear of the train. Guard Proudfoot overlooked this rule and omitted to do so. There were several qualified flagmen in the ballast gang which consisted of about 18 men.

The ballast train was stopped in the first place with the engine at the south end of Waste Bank tunnel, but as the work progressed, it was moved on two or three times, and at the moment of collision the engine was about the middle of the tunnel.

At 9.58 a.m. the 5.10 a.m. mineral train Skipton to Carlisle arrived at Culgaith, and as the passenger train was due immediately after, signalman Wileman shunted the mineral train on to the up road to allow the passenger to pass.

Flagman Taylor had not been recalled to his train at this time. He observed this shunting movement taking place at Culgaith Station, and, realising that the passenger train must be approaching shortly, decided to return to his train without waiting to be recalled, and to move the train on to clear the road. He picked up the three detonators which he had put down, and walked back to his train, also picking up en route the three detonators he had put down at the north end of Culgaith tunnel. When he reached the train, he called to ganger Watson that it was time they were off, and the latter immediately ordered his gang to put the doors up and get on board.

The passenger train " entering section " signal was received at Culgaith box at 10.7 a.m. and signalman Wileman asked signalman Kirkpatrick at Langwathby if he could see any sign of the ballast train, but the reply was in the negative. The block instrument was correctly showing " train on line." Wileman kept his home and starting signals at danger, with a lever collar on the starting signal lever, until the passenger train approached the former and whistled for the signal, when it was pulled off to allow the train to enter the station under Rule 40 (See Appendix). It is questionable whether the home signal was kept at danger for long enough to comply strictly with Rule 40 and thus to convey to the driver the warning intended by this rule.

The train arrived at the platform at 10.9 a.m., i.e. five minutes late. There were three passengers embarking and a few parcels and milk vans to be dealt with at the station. Guard Brogden was riding in the second vehicle, the 6-wheeled brake van, and, after handling the parcels with porter Nixon, the latter closed the doors, saw to the handles, and gave the " all right " signal to the guard. Nixon noticed that at this moment the starting signal was at danger but he did not observe it later. Guard Brogden looked back along the train to see that all doors were closed, and then gave " right away " to the driver without looking at the starting signal, got into his van, and started to arrange his parcels for the next station, while the train moved off.

As soon as the train had come to a stand signalman Wileman opened the gates of the level crossing behind the train to allow a road vehicle to pass, and he was then booking the train in his register, when fireman Armstrong, of the shunted mineral train, who was waiting in the box in accordance with Rule 55, saw the passenger train start away against the starting signal at danger. He immediately drew Wileman's attention to this, and for a moment they thought the driver might be only drawing up to the starting signal, which is some 130 yards beyond the platform, but the train went on past the signal and Wileman could do nothing but signal to Langwathby " vehicles running away on right line " at the same time stating what had happened and inquiring if the ballast train was in sight of Langwathby. This was of course not the case.

Part of the damaged passenger train pictured just outside the Culgaith end of Waste Bank tunnel having been removed. The northern entrance of the tunnel is out of shot to the left of the camera.

This picture shows the damaged passenger train guard's van presumably within the tunnel rather than at night. The tunnel is only 164 yards in length but does have a curve.

Two views of the locomotive involved in the Culgaith incident, No. 5971 CROXTETH, both in LMS livery.

LONDON ELECTRIFICATION SCHEMES

FROM THE RAILWAY GAZETTE, 14TH AUGUST 1914

LONDON & NORTH WESTERN RAILWAY

With the delivery of three out of the four three-coach trains, the London & North Western electric railway service between Willesden and Earl's Court, which has up to the present time been operated by District Railway stock, will be operated by these specially designed trains, which are illustrated in the present issue.

These trains, which consist of motor coach, driving trailer, and trailer, have been built by the Metropolitan Carriage, Wagon & Finance Co., Ltd., and, in details at least, present certain special features. They are of the end-door type with through communication, 56 ft. 11 in. over headstocks.

The motor cars are carried on two four-wheel bogies of 8 ft. 9 in. wheelbase and 37 ft. 6 in. centres. The driving trailer and trailer coaches have a wheelbase of 9 ft., and are carried on 38 ft. centres. The wheels are 3 ft. 5 in. diameter in all cases. The underframes of the coaches are of steel and the floor is plated underneath for purposes of fire resistance. The driving end of the motor car is built entirely of steel and the side panels are also constructed of that material.

The interior finish is in polished mahogany. The electrical control Switch gear and the driving gear is all contained in the cab, there being none underneath the coach. As is customary in installations of this character, the control is on the multiple-unit system with the usual automatic features. The motor coaches will be equipped with four motors of 250 h.p., each carried on bogie trucks of specially strong design as may be noted from the photographs.

The lighting of the cars is effected by central and side bracket holophane electric lamps, the central lamps containing two 16 C.D. metallic filament lamps, and the side brackets one 16 c.p. The seating accommodation varies with the type of car, the motor coach containing smoking accommodation for 40 third-class passengers, the trailer cars accommodation for 20 first-class non-smoking and 16 first-class smoking, and the driving trailer being a smoking compartment for 60 third-class passengers. Two of the four trains are fitted with air compressors of the Westinghouse Company's make, and the other two with

Driving Trailer.

First Class interior.

Third Class interior.

the Knorr Brake Company's rotary slide valve compressors.

The main electrical equipment for these trains is being furnished by Messrs. Siemens Bros. Dynamo Works, Ltd. The three trains delivered will be put into service between Willesden and Earl's Court as soon as the necessary trial runs have been carried out. Pending the completion of the Company's own power station at Stonebridge Park, current for these services is being furnished by sub-stations of the Great Western and District railways.

PROGRESS ON EUSTON–WATFORD LINE

While the Willesden–Earl's Court has been merely an equipment of existing tracks, the determination to furnish electric services between Euston and Broad Street and Watford has necessitated the provision of entirely new permanent way between Chalk Farm and Watford. As between Willesden and Watford, the new permanent way has been completed for some time past and the new local stations opened for traffic, and operated by steam trains.

Meanwhile, the work of equipping these lines for electric traction is proceeding satisfactorily, a large mileage of third and fourth rails having been placed in position, the positive rails being laid outside the track rails, and the negative rails between. The conductor rails are of Special low carbon soft steel, weighing 105 lb. per yard, and the composition being carbon 0-044, manganese 0-139, silicon 0-030, phosphorus 0-011, sulphur 0-029, and nickel 0-255 per cent. The volume resistivity of the rails is approximately six and a half times that of copper and they are being manufactured by the Cargo Fleet Iron & Steel Company. The conductor rails are supported on insulators of the usual pattern attached to the sleepers by malleable iron clips, special anchor insulators being provided at intervals to prevent creeping of the rails. These insulators have been supplied by Messrs. Doulton & Co. Each rail joint is bonded by four copper bonds of the flexible strip type supplied by the Forest City Electric Services Supply Company. The bonds are of solid drop forged heads and are fixed into the rails by hydraulic pressure. The trailing ramps are of cast iron and the leading ramps on the through lines are of forged steel, and in some cases, as on sidings, of special cast steel. The jumper cables for connecting the several sections of the rails together at crossovers, etc., have been supplied by Messrs. W. T. Henley's Telegraph Works Company, and have been laid solid in bitumenised fibre troughs and fitted with the Cortez Leigh patent sealing terminal, which forms an effective seal to the installation of the cable, and prevents damage arising from mechanical shocks and vibration.

Motor car.

L&NWR AND ALLIED LINES NOW BEING ELECTRIFIED

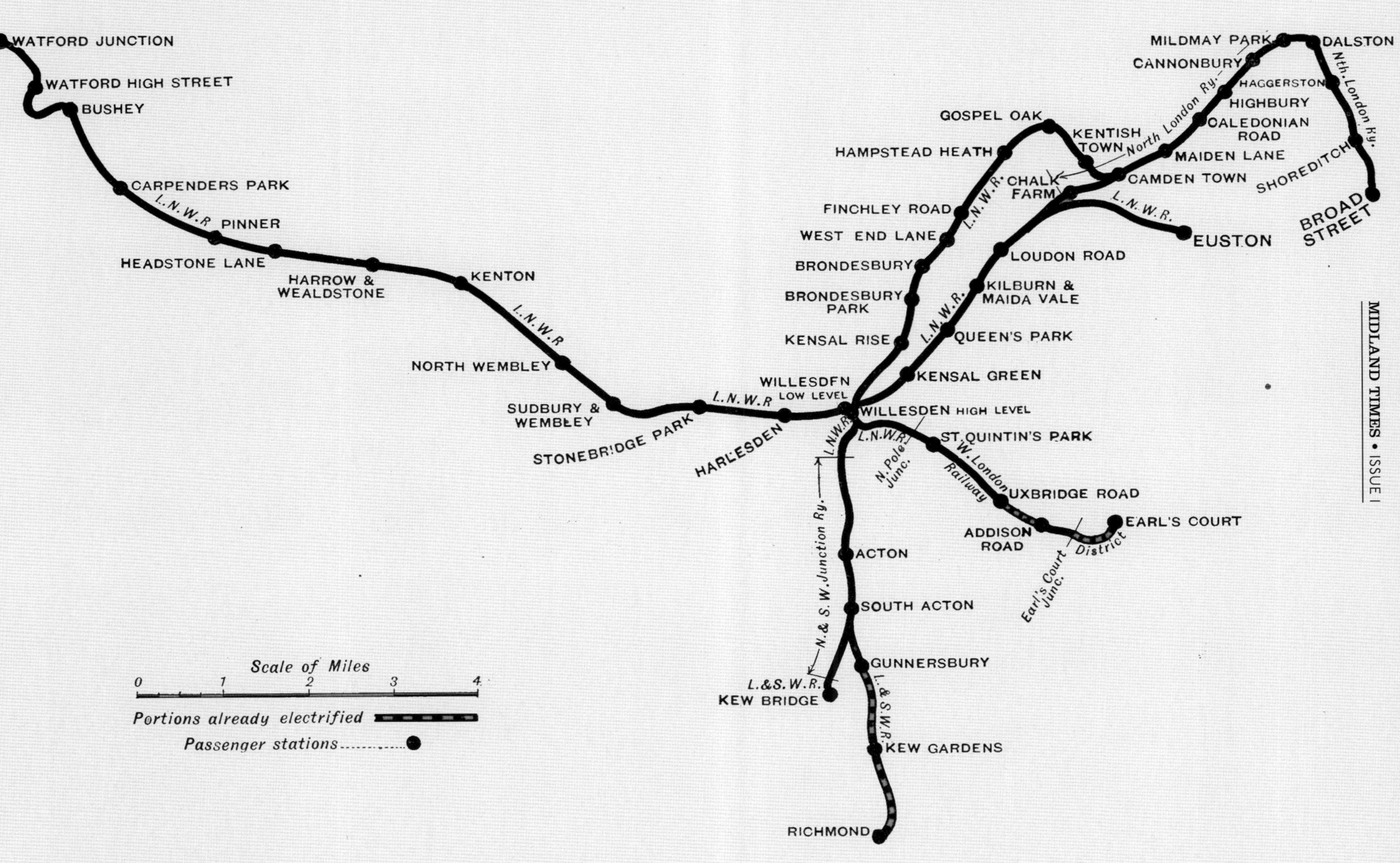

Messrs. Henley are also providing the feeder and sectionalising Switch pillars.

This work is going steadily forward, and work is also in an advanced stage on the section of new line between Queen's Park Station and Willesden, the object being to get this work completed by the time the extension of the Baker Street & Waterloo Railway from Paddington to Queen's Park is finished and to inaugurate at the earliest possible date through services between the Tube Railway system and Watford. For this purpose Queen's Park Station, where the Tube Railway will come to the surface and effect a junction with the electric tracks of the London & North Western Railway, is being reconstructed, and provision is being made for handling a much larger traffic. The new station will be provided with six platforms and adequate car shed accommodation for the joint stock which is being constructed for the through services from Watford over the Baker Street & Waterloo system.

Under the new arrangement the Queen's Park goods yard will also have to be reconstructed. A new station is being built at Kensal Green on the east side of the Kensal Green tunnel, and two new single track tunnels of 16 ft. 4 in. internal diameter are being constructed on the shield system and are probably, with those at Chalk Farm referred to in what follows, the largest diameter tunnels yet constructed on the shield method. The tunnel for the up electric services is finished, and that for down electric trains is making good progress. From the west face of the tunnel to the new local station which has been built at Willesden, the work presents no exceptional features, and it is hoped that by the end of the present year trains will be running from the Bakerloo system to Watford.

A great deal of work yet remains to be done between Chalk Farm and Queens Park including the driving of two long tunnels through the London clay at Primrose Hill. The tunnels which are of 16 ft. 4 in. diameter, are each about a mile in length, and it is not intended to commence driving them until September next, they are not likely to be complete until the spring of 1916. From the western end of these tunnels to Queens Park, the permanent way will present no special features, the most important work perhaps being the reconstruction of Kilburn goods yard.

East of the Primrose Hill tunnels, a difficult and costly piece of work is, however, being undertaken, and the whole of the existing permanent way, with the exception of the down fast track, is to be reconstructed. The Chalk Farm Station of the London & North Western Railway is being demolished, the intention being to make provision for the whole of the local traffic at Chalk Farm at the North London Station of that name.

Between Euston and Chalk Farm the existing slow line tracks are being employed for the electric services, the two new tracks beginning at a point a short distance west of Chalk Farm. The object of the track reconstruction scheme is not only to provide the new lines for the electric services but to obviate the crossing on the level of the down and up slow lines over the up fast line, which is a disadvantage of the present system of working. To achieve this result, the down electric line will be carried in tunnel from a point near the new steel lattice girder bridge which is being built over the Regent's Park Road, under all the existing lines coming to the surface at Loudoun Road Station. The up electric line will also be carried in tunnel from Loudoun Road to Chalk Farm and will make a junction with the ordinary up slow line underneath a new fly-over bridge which is being constructed to carry the up fast line. It is also necessary in the reconstruction scheme to make provision for the junction with the North London Railway lines by which the electric trains will reach Broad Street. This will be an underground junction.

In addition to the new electric tracks an empty carriage road, which will come off the up slow line and which will be carried in a tube to a point near Gloucester Road, is to be constructed, the object being to enable empty trains to be brought from the Willesden carriage shed into Euston Station without fouling the passenger line. It may be pointed out that a special line was provided some years ago in the reconstruction between Euston and Chalk Farm for locomotives going from Chalk Farm engine sheds into Euston.

L&NWR 3-car Electric Train.

STONEBRIDGE PARK POWER STATION

Excellent progress is being made with the erection and equipment of the power station at Stonebridge Park. Several of the twenty Babcock & Wilcox boilers are now in position. The boilers are being arranged in two rows with a separate chimney stack for each row, and these stacks, which will each be of 14 ft. internal diameter and 240 ft. in height are now about half finished.

Fairly rapid progress has also been made with the coal and ash handling plant and a start will soon be made with the installation of the five Westinghouse turbines driving Siemens three phase alternators to generate current at 11,000 volts, which will form the first section of the plant; the power is house being designed of sufficient area, however, to allow the plant capacity to be more than duplicated.

Work is also in hand on the eleven sub-Stations at Stonebridge Park, Bushey, Headstone Lane, Kenton, Willesden Junction, Queen's Park, West End Lane, Chalk Farm, Dalston, Broad Street, and South Hackney. The sub-station plant, which will consist in each case of three rotary converters, nine single-phase static transformers, a large storage battery, and an automatic reversible booster, is being supplied by the British Thomson Houston Company. The converting units will be 750 kW and 1,000 kW, with a large overload capacity. The high-tension cables will be of the British Insulated & Helsby Company's three core paper, insulated, lead covered and of the armoured type.

The trains for the main line services, which are now under construction, are being built by the Metropolitan Carriage, Wagon & Finance Company, who are supplying the motor coaches, and by the London & North Western Company's works at Wolverton, where the trailers are being constructed. The electrical equipment is being carried out under the direction of Mr. E. F. C. Trench, Chief Engineer of the Company.

PERFORMANCE AND EFFICIENCY TESTS LMR 4-8-4 DIESEL-MECHANICAL LOCOMOTIVE

FELL SYSTEM, LOCO No. 10100 – C.M.&E.E. DEPARTMENT, DERBY – 1955

10100 passing Tebay with a test train for Crewe having traversed the Settle & Carlisle line.
This was one of a series of trials which had commenced on 25th April 1955 and involved a Dynamometer car (seen here as the first vehicle) as well as the LMS Mobile Test Unit. On one of these tests the 18 miles from Appleby to Ais Gill were covered in just 25 minutes with 389 tons in tow. Using full throttle, 1,900 h.p. was measured on the drawbar at 44 m.p.h., a commendable achievement considering a theoretical 100 h.p. was all that was left to move the locomotive itself as well as cover traction losses. The starting tractive effort was recorded at 29,400 lbs or just over 13 tons.

INTRODUCTION

The Fell Diesel-Mechanical Locomotive was built in the LMR Derby Locomotive Works in 1950. The design was based on a new principle in diesel power application and transmission, as proposed by Lt. Col. Fell, and the locomotive was jointly developed by Mr. H. G. Ivatt, the LMR Chief Mechanical Engineer at the time.

The aim of the design was to reduce transmission losses as compared with diesel electric traction and also to reproduce the performance characteristics of the steam locomotive as far as practicable. For this purpose, the four main engines are supercharged and governed to produce substantially constant horsepower over the speed range. The output of the engines is combined in a specially designed gearbox.

The primary object of the tests was to establish the drawbar power characteristics and corresponding fuel consumption over the whole working range of the locomotive. Additional tests were carried out to investigate particular aspects of design.

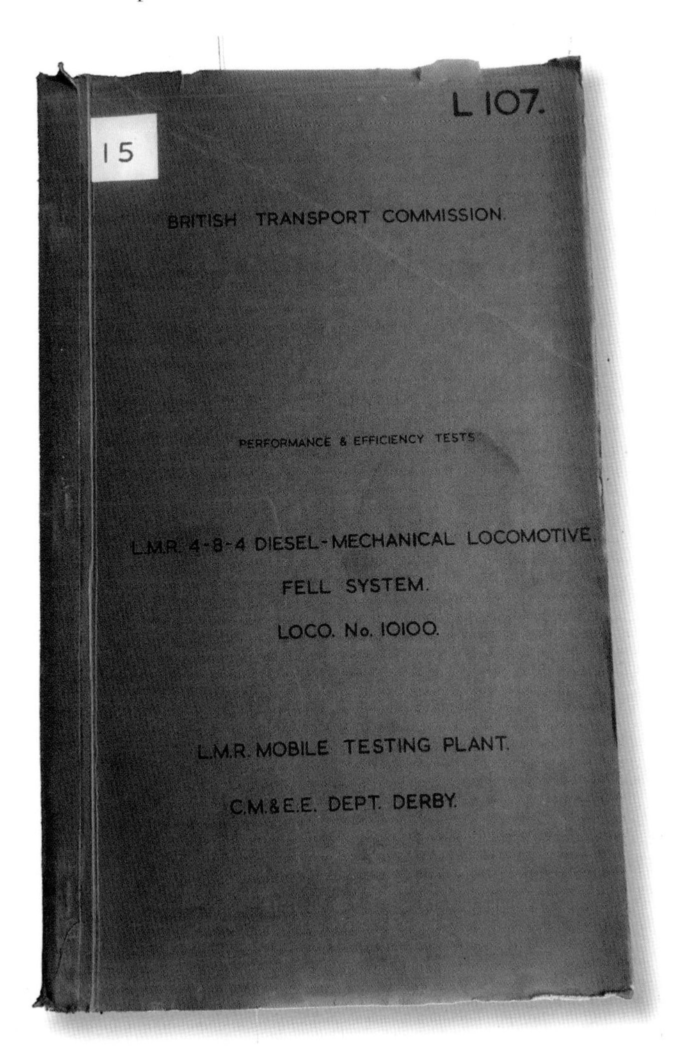

As received, the performance of the locomotive, particularly in the low speed range, was not as good as anticipated. The first part of the tests was, therefore, occupied in tuning the locomotive and a substantial improvement was achieved.

THE LOCOMOTIVE

The motive power consists of four Diesel Engines, each rated at a maximum of 510 b.h.p. and supercharged by two separately driven blowers. The blowers, together with the other auxiliaries, are driven by two A.E.C. Diesel Engines, each with a maximum output of 150 b.h.p.

In order to achieve the constant horse power required, with the corresponding falling torque characteristic, the main engine governors are designed to reduce the amount of fuel injected as the speed increases. Also the blower capacity is limited so that the boost pressure falls from approximately 10 lb/sq.in at a main engine speed of 500 r.p.m. to 2 lb/sq.in at 1,500 r.p.m. by automatic adjustment of the auxiliary engine speed.

The transmission from the main engines to the road wheels is by means of a gearbox incorporating differential gears, each engine being connected to the gearbox through a scoop controlled fluid coupling. Each of the two primary differentials combines the output of the pair of engines at one end of the locomotive and the secondary differential couples the outputs from these primary differentials and passes the torque through fixed ratio gearing to the intermediate driving axles of the locomotive. The purpose of the gearbox is to extend the speed range of the locomotive beyond that which could be obtained by a direct drive from a single power unit.

Considering the differentials alone, their effect is to decrease the gear ratio as successive engines are engaged. Thus, with one engine driving the ratio is 4:1, with two engines 2:1, with three engines 1.1/3:1, and with four engines 1:1. In each case the same total torque is maintained, but the speed range is extended. For example, as the second engines torque is added to that of the first, the gear ratio is halved, thus the same torque at the road wheels is maintained at the higher road speed. In order to provide a higher initial tractive effort than would be obtained from this basic arrangement, the input gear ratio for one engine has been raised thus making it the starting engine. The transfer gear ratio between one primary and the secondary differential has also been increased.

Starting from rest is carried out on one engine, which alone will only produce a road speed of approximately 14 m.p.h. at maximum engine r.p.m. (i.e. nominally 1,500). The second engine, is, therefore, normally engaged well before the first has reached maximum revs and will begin to drive as soon as the torques of the two engines are in balance at the differential. Thereafter, the output of the two engines is combined with a decreased gear ratio, this permitting a maximum road speed of approximately 30 m.p.h. at maximum engine revs. In a similar manner a top speed of 51 m.p.h. is reached with three engines only and 72 m.p.h. with all four engines driving. The road speed at which each engine should be engaged is indicated on the driver's speedometer.

The manual controls comprise: (1) The regulator which is connected to all main engine governors and provides an overriding control on the fuel feed. Thus, proportions of full power output can be obtained, still governed to produce the required falling torque characteristic; (2) Four Scoop Control Levers, which control the operation of the scoop tubes on the four fluid couplings; (3) The Reverse Lever.

When a Scoop Control Lever is placed in the scoop withdrawn position, the regulator control on the main engine concerned is uncoupled, so that whatever the setting of the Regulator, the engine remains in the slow running position. All controls are duplicated, so that the locomotive can be driven from either end.

An engine diagram (Fig. 1 below) and photographs of the locomotive are shown on the right. Leading dimensions and data are given in Table No. 1.

The locomotive had run 24,000 miles since receiving a heavy casual repair.

ABOVE RIGHT: The official view of the 'Fell' as built with 4-8-4 (2-D-2) wheel arrangement.

BELOW RIGHT: 10100 pictured at the International Railway Congress that was held at Willesden from 2nd May to 4th June 1954. Basic technical details are given on the information board which states a maximum speed of 75 m.p.h. and a weight of 120 tons. The livery was gloss black and possibly for the exhibition, the outside cranks were painted bright red.

Fig. 1.

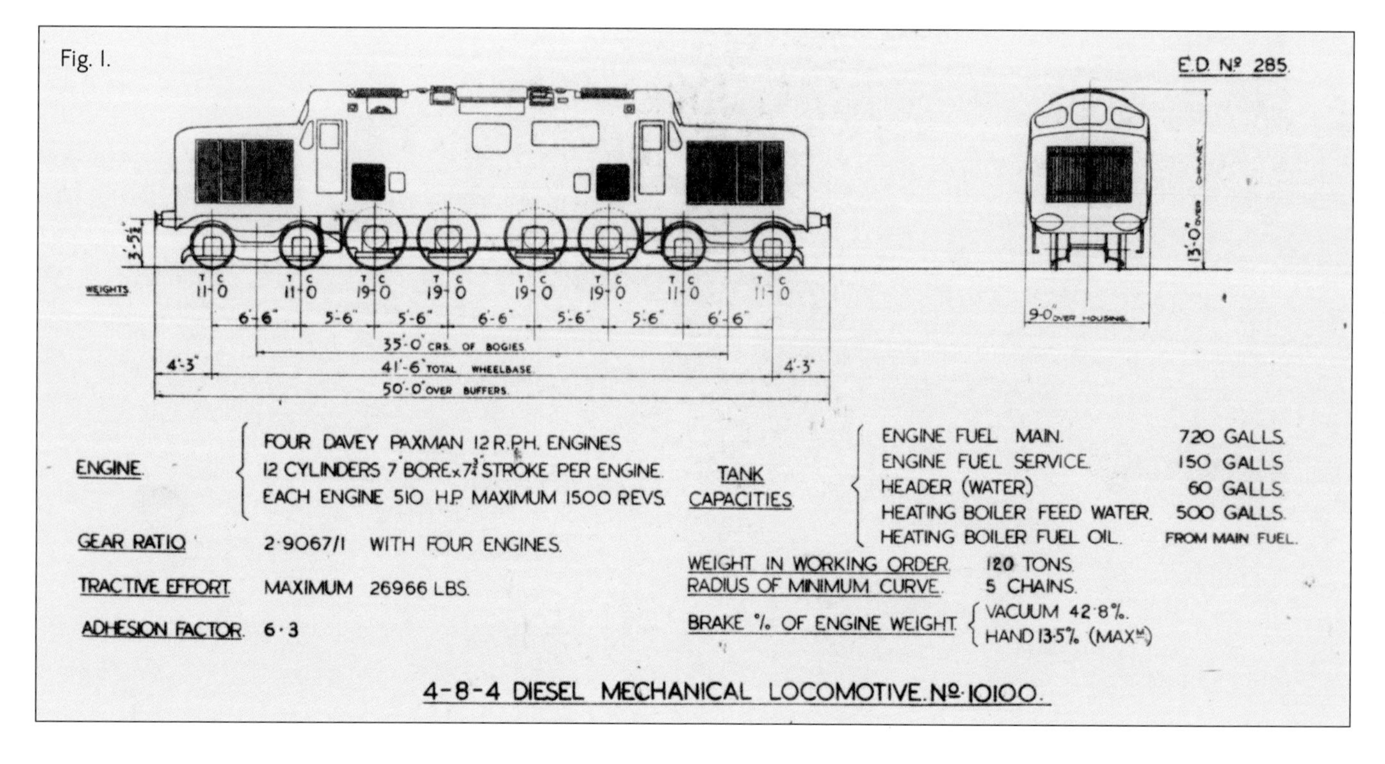

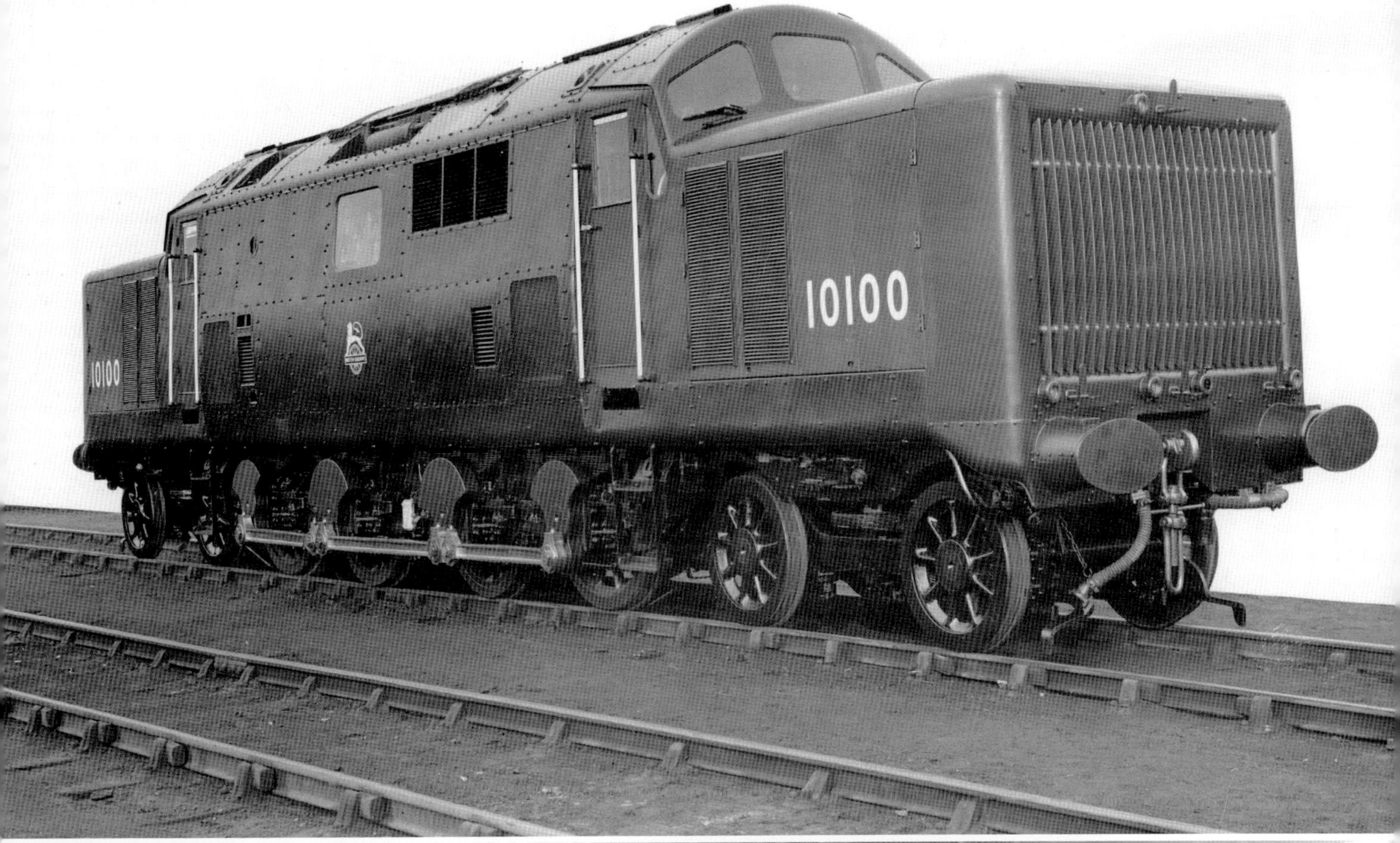
10100
10100

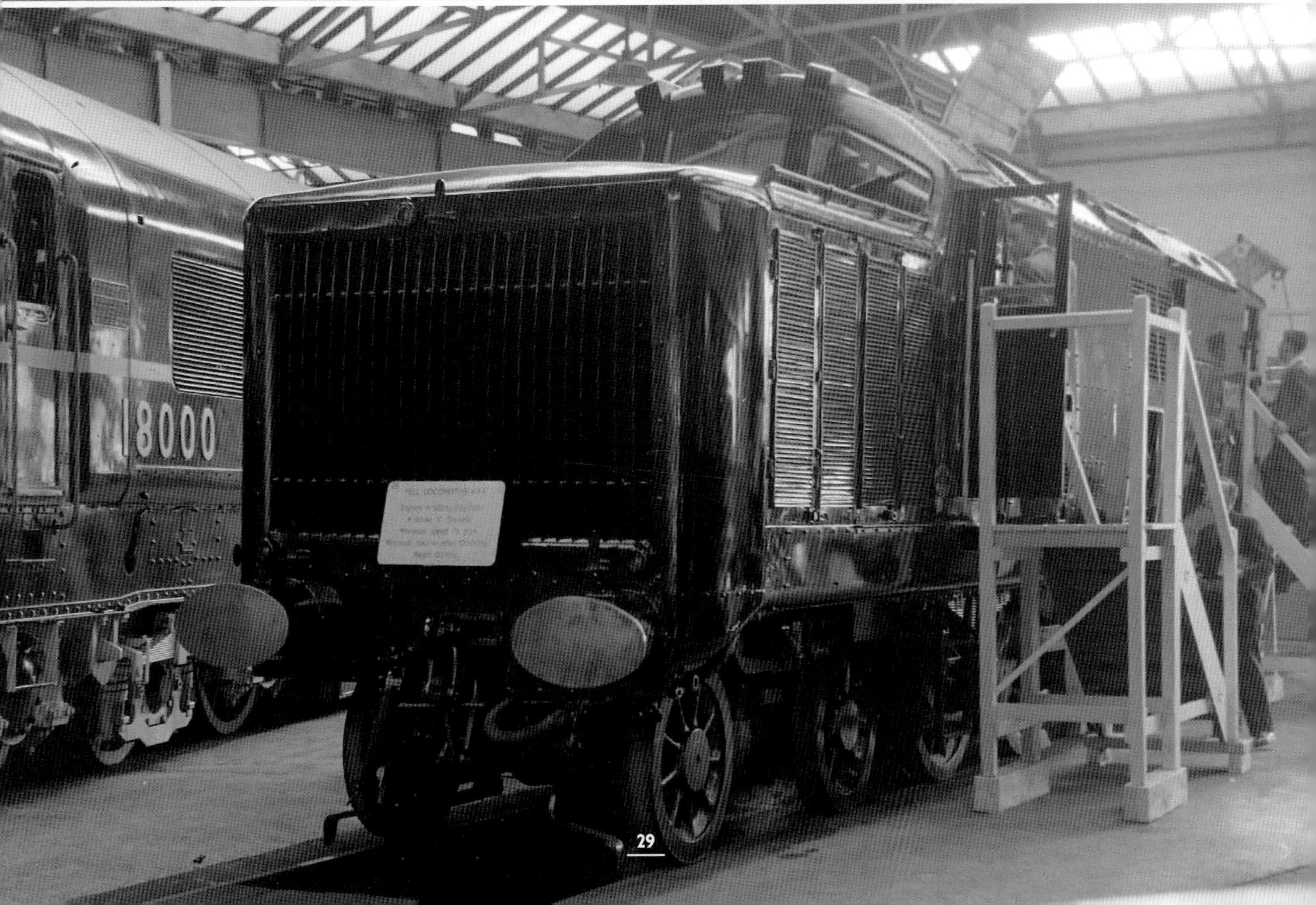
18000
FELL LOCOMOTIVE

TABLE I

L.M.R. 4-8-4 DIESEL MECHANICAL LOCOMOTIVE – FELL SYSTEM
LOCOMOTIVE No. 10100

LOCOMOTIVE DATA	
Type	2-D-2 (4-8-4)
Weight in running order	120 tons
Tractive Effort (maximum)	26,966 lbs
Rigid Wheelbase	17ft 6in
Total Wheelbase	41ft 6in
Driving Wheels – Diameter	4ft 3in
Bogie Wheels – Diameter	3ft 0in
Width overall	9ft 0in
Length overall	50ft 0in
Height overall	13ft 0in
Minimum curve negotiable	5 chains
Maximum permitted speed	72 m.p.h.
Adhesion Factor	6.3
Capacity – Fuel Tanks	
Two main tanks	360 gallons each
Two service tanks	66 gallons each
Total	852 gallons
Capacity – Lubricating Oil	
Four Main Engines	40 gallons each
Two Auxiliary Engines	20 gallons each
Gearbox	80 gallons
Four Fluid Couplings	26 gallons each
Capacity	
Cooling Water Header Tanks (two)	30 gallons each
Boiler Water Tanks (two)	250 gallons each
Brakes (% of locomotive weight)	Vacuum 42.8%
	Hand 13.5% max.
Sanding	Gravity

POWER EQUIPMENT
Four 12-cylinder Diesel Engines (supercharged) Davey Paxman and Company Limited, Series R.P.H. Maximum Horsepower: 510 b.h.p. @ 1,000 r.p.m. Maximum permitted speed: 1,500 r.p.m. (A&B engines adjacent to No. 1 Cab) (C&D engines adjacent to No. 2 Cab) Cylinder Bore and Stroke – 7in x 7¾in
Two 6-cylinder Diesel Engines (auxiliary) A.E.C. Supercharged Maximum Horsepower: 150 b.h.p. @ 1,300 r.p.m. Maximum permitted speed: 1,800 r.p.m. Cylinder Bore and Stroke – 4.72in x 5.51in
TRANSMISSION
Fell Gearbox – incorporating differential gears as a means of grouping together the output of 1, 2, 3 or 4 propelling engines
Gear Ratios – 1 engine at 1,500 r.p.m. – 16.52 to 1 2 engines at 1,500 r.p.m. – 7.19 to 1 3 engines at 1,500 r.p.m. – 4.30 to 1 4 engines at 1,500 r.p.m. – 3.07 to 1
Reverse Gear and Final Drive – fitted in lower section of gearbox, the reverse mechanism consisting of sliding dog clutches, vacuum-operated
AUXILIARIES
Superchargers (2) Holmes Connersville Roots type blowers driven by the auxiliary engines. Size 10in x 13½in.
Boost Air Intercoolers (2) Spiral tube type roof radiators. Serck Heat Exchangers. Water circulation pump driven by auxiliary engines.
Lighting and Starting – 24 volt Lead Acid Batteries 1-1.3 kW generator driven from each auxiliary engine
Exhausters (2) Westinghouse 3V72 type. Belt-driven from auxiliary engines
Radiator Fans, Engine Water Pumps, Gearbox Oil Pumps – mechanically driven from auxiliary engines
Carriage Warming Apparatus – 1 Sentinel type Boiler Maximum output – 1,000 lbs/hour Maximum Boiler Pressure – 100 lbs/sq.in. Automatic Water Feed and Oil Firing equipment by Laidlow Drew
Brake System – Vacuum
Control System – Mechanical with vacuum relays

TABLE 2

ANALYSES OF SHELL GAS OIL SUPPLIED TO BRITISH RAILWAYS AT KINGMOOR, CARLISLE, FOR DYNAMOMETER CAR TESTS OF FELL LOCOMOTIVE No. 10100

	Rail Tank* Car No. 2382	Rail Tank Car No. 55	Rail Tank Car No. 3817
Specific Gravity, 60°/60°F	0.830	0.830	0.830
Viscosity, Kinematic at 100°F (centistokes)	–	2.55	2.55
Carbon Residue (% wt.) (Conradson)	–	0.02	0.02
Distillation			
Initial Boiling Point (°C)		177	178
10% volume recovered at (°C)		205	205
20% volume recovered at (°C)		215	215
30% volume recovered at (°C)		226.5	227.5
40% volume recovered at (°C)		240	241.5
50% volume recovered at (°C)		255.5	257.5
60% volume recovered at (°C)		273.5	275
70% volume recovered at (°C)		292.5	293.5
80% volume recovered at (°C)		314	314
90% volume recovered at (°C)		344	342.5
Recovery at 350°C (% volume)		92	92
Total recovery (% volume)		96	95.5
Residue and Loss		4	4.5
Flash Point (°F)	–	156	154
Calorific Value (Gross, B.Th.U./lb.	–	19,540	19,480
Ash (% wt.)	–	Less than 0.01	Less than 0.01
Sulphur (% wt.)	–	0.74	0.77
Acidity (Inorganic)	–	Nil	Nil
Corrosion (Copper Strip) at 212°F	–	Negative	Negative
Cetane Number	–	53	53

*Rail Tank Cars Nos. 2382 and 55 were filled simultaneously from the same storage tank at Heysham Refinery. Since the specific gravities were identical full analyses were only made on one sample.

NATURE OF THE TESTS

The first part of the tests was devoted to checking the effect of adjustments to the fuel control system, particularly at low speeds. After a substantial improvement had been effected, a comprehensive series of tests was made, primarily to establish drawbar tractive effort characteristics and corresponding fuel consumptions. In order to cover the operating range of the locomotive, tests were conducted in four regular positions approximately corresponding to:

1. ¼ power output, engines normally aspirated.
2. ½ power output, engines normally aspirated.
3. ¾ power output, engines supercharged.
4. Full power output.

The locomotive was tested in each of the regulator positions with 1, 2, 3 and 4 engines. The tests covered a speed range of 4-70 m.p.h., the maximum speed of the locomotive under power being 72 m.p.h.

Since provision is made for the auxiliary engines to run at idling speed when the regulator is partially closed, the engines are not supercharged for the lower power outputs.

In addition to measuring drawbar power and fuel consumption, particulars of engines revs boost air pressure and temperature, fluid coupling oil temperatures etc., were recorded during each test.

Finally, special tests were made to investigate particular aspects of performance as follows:

1. Static tests to determine the starting drawbar tractive effort.
2. A test of over one hour duration at maximum power to determine the rise in gearbox oil temperature under such conditions.
3. A test under service conditions with a Class 6P full load of 390 tons, during which carriage warming was in operation. The steam flow was measured together with air temperature and C.W. steam pressure at various points in the train.
4. A test of 30 minutes duration at 31 m.p.h. where the engine torque and boost pressure were approaching maximum values.

METHOD OF TESTING

The tests were conducted between Carlisle and Skipton, London Midland Region. The L.M.R. Mobile Test Plant was employed, and comprised No. 3 Dynamometer Car with Amsler equipment, and three electrically braked Mobile Testing Units which enabled tests to be conducted at constant speeds.

The daily point-to-point running times were such as to permit a succession of constant speed tests covering the speed range of the locomotive, each test being of 10-15 minutes duration. In the lower speed range, i.e. 10-30 m.p.h., test speeds were selected as required to establish the irregularly shaped Drawbar Characteristics. At higher

speeds with four engines engaged, tests were made in speed increments of approximately 5 m.p.h.

During each test, Drawbar T.E., Drawbar H.P. and speed, etc. were measured in the dynamometer car, whilst observers on the locomotive recorded fuel consumption and the various temperatures and pressures. Telephonic communication together with pens on the dynamometer car instrument table, which were electrically operated from the locomotive, provided the means for synchronising test readings and co-ordinating the work of the various operators.

Four fuel oil flow meters were employed; one for each of the two A.E.C. auxiliary engines and one to measure the inclusive fuel consumption of each pair of main engines and their associated auxiliary engine. The displacement type meters were calibrated before and after the trials and all engine revolution indicators, pyrometers, etc., were also calibrated. The fuel oil temperature was recorded and used to correct the measured oil flow to an equivalent consumption at a standard temperature of 60°F.

With any particular number of engines engaged, fuel consumption varied with speed. A constant speed testing technique was, therefore, essential for the accurate determination of fuel consumption characteristics.

In the early stages of the tests a deficiency in power output was observed, particularly at low speeds. This had not hitherto been detected since the speeds involved were passed through during acceleration in service running without attracting attention. Owing to the critical nature of the adjustments, the subsequent investigation required a large number of tests at constant speeds in increments of 1 m.p.h.

FUEL OIL

The locomotive was fuelled from Shell Rail Tank cars and analyses of the fuel oil are given in Table No. 2 (page 31). The average Gross Calorific Value was 19,250 B.Th.U./lb.

LEFT: 10100 partially undressed with the complexities associated with the four separate engines visible. Below the framing the final drive was by a train of gears. Reversing was by vacuum operated sliding dog clutches, so the engine could run equally well in either direction.

METHOD OF PRESENTATION OF RESULTS

The report has a series of graphs (36 in total) relating to the performance and efficiency of the locomotive which are divided into three parts:

PART 1 Drawbar Characteristics and Fuel Consumptions
PART 2 Diesel Engine and Transmission Characteristics
PART 3 Operating Characteristics

As stated under the heading 'Nature of the tests', the main tests for measurement of Drawbar Power and Fuel Consumption were carried out for four regulator positions:

REGULATOR POSITION 1

This is the 'Latch' position which is a fixed stop on the regulator quadrant and is the lowest setting of the regulator for working under power.

REGULATOR POSITION 2

This position was selected as being the maximum regulator opening at which the engines are not boosted, and gave approximately half power output.

REGULATOR POSITION 3

This position was chosen in a similar manner to give the minimum regulator opening at which the engines are boosted. It was located 0.04 in. from position 2 and gave approximately ¾ power output.

REGULATOR POSITION 4

Fully open to give maximum power output.

Due to space constraints only the final graph (Fig. 36) is shown (page 34), this forms part of the conclusions to the trial.

CONCLUSIONS

MECHANICAL PERFORMANCE

During the course of the tests, the locomotive ran a total of approximately 3,300 miles free from mechanical defect. Working for comparatively long periods at maximum power and with maximum boost had no adverse effects other than the rather high temperature produced in C engine's fluid coupling. The gearbox, which is the outstanding feature of the design, was trouble free throughout, and the temperature rise experienced under the most arduous working conditions was considered to be quite permissible.

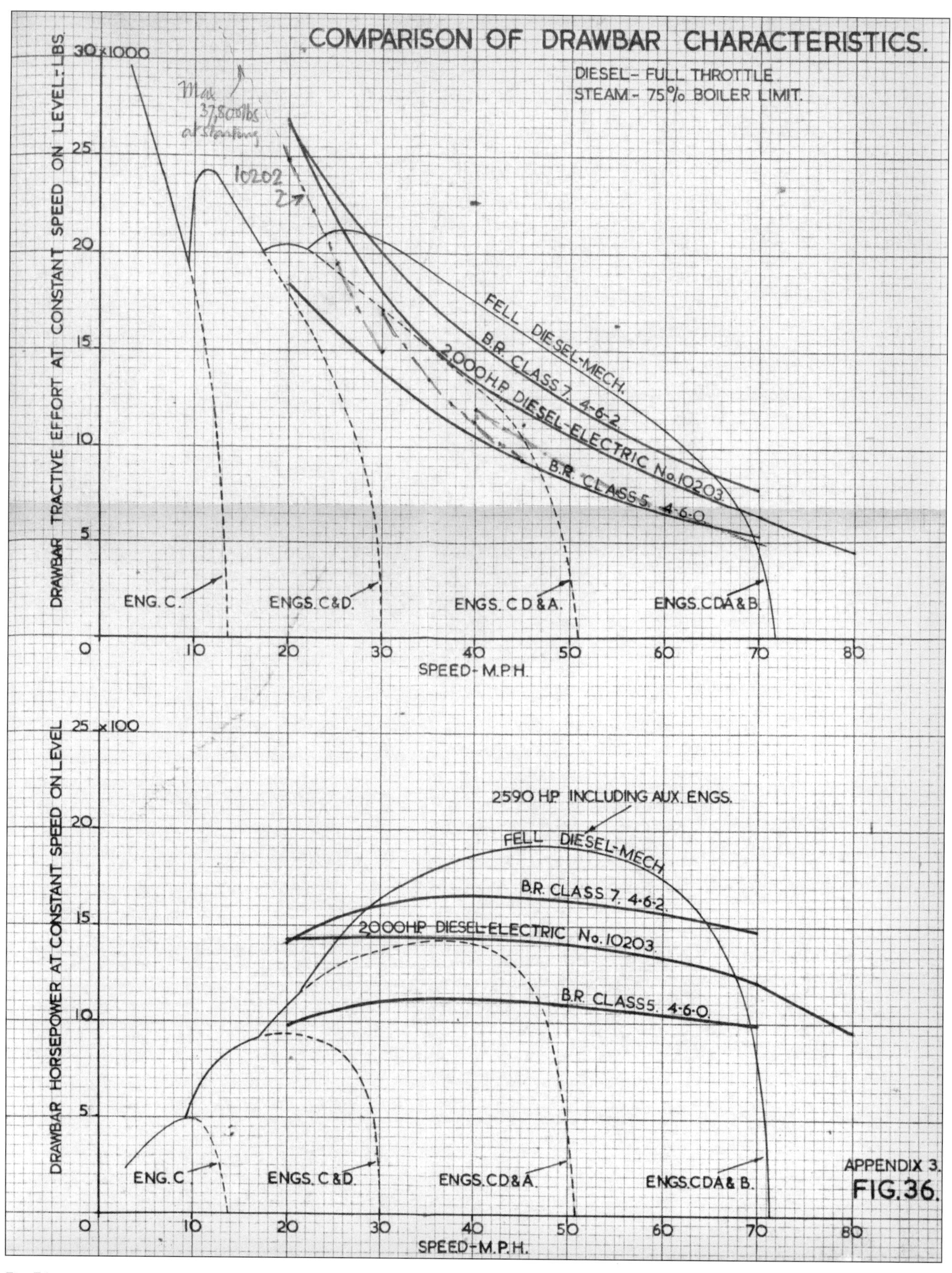
COMPARISON OF DRAWBAR CHARACTERISTICS.
DIESEL – FULL THROTTLE
STEAM – 75% BOILER LIMIT.
DRAWBAR TRACTIVE EFFORT AT CONSTANT SPEED ON LEVEL-LBS.
30 x1000
10202
FELL DIESEL-MECH.
B.R. CLASS 7. 4-6-2.
2000 H.P. DIESEL-ELECTRIC No. 10203.
B.R. CLASS 5. 4-6-0.
ENG. C.
ENGS. C&D.
ENGS. C D & A.
ENGS. C D A & B.
SPEED-M.P.H.
DRAWBAR HORSEPOWER AT CONSTANT SPEED ON LEVEL
25 x100
2590 H.P. INCLUDING AUX. ENGS.
FELL DIESEL-MECH
B.R. CLASS 7. 4-6-2.
2000 H.P. DIESEL-ELECTRIC No. 10203.
B.R. CLASS 5. 4-6-0.
ENG. C
ENGS. C & D.
ENGS. C D & A.
ENGS. C D A & B.
APPENDIX 3.
FIG. 36.
SPEED-M.P.H.

Fig. 36.

POWER AND FUEL CONSUMPTION

The maximum power development of the Fell locomotive is compared with that of a 2,000 h.p. Diesel Electric (No. 10203) and two B.R. Standard steam locomotives on Fig. 36 (left).

It should be noted that the four 'Fell' main engines are normally rated at 510 b.h.p. but for reasons given in the report, were actually producing 560 b.h.p. (maximum) during tests, making a total of 2,240 b.h.p. Two 150 h.p. engines provide the power for auxiliaries and for blowing the main engines. Comparing the 'Fell' diesel mechanical and the 2,000 h.p. diesel electric locomotive, the following features emerge:

1. The comparatively 'flat' d.b.h.p.-speed characteristic of the steam locomotive is more faithfully reproduced by the diesel electric than the 'Fell' locomotive.
2. At 50 m.p.h. 93% of the Fell b.h.p. is obtained at the rail compared with 84% in the case of the diesel electric, after deducting the power required for auxiliaries.
3. In spite of the higher transmission efficiency of the Fell locomotive, its overall efficiency in terms of fuel per d.b.h.p. hour is barely as good as the diesel electric. The specific fuel consumption at 50 m.p.h. and maximum power are 0.56 and 0.54 lb per d.b.h.p. hour for the Fell and diesel electric locomotive respectively.

DESIGN FEATURES

There are certain other features connected with the design and performance of the locomotive which call for comment.

(1) CAB AND ENGINE ROOM TEMPERATURE

Maximum temperatures of 107°F and 130°F were recorded in the trailing locomotive cab and engine room respectively during the trials. The tests were conducted during April and May 1955 when the ambient temperatures were of the order of 44-64°F.

(2) BOOST AIR SYSTEM

It has been shown in the report that individual engines do not get their designed boost under all conditions of working. This could be overcome by separate boost air supply to each engine whereas at present one blower supplies a pair of engines.

Another feature of the boost system is the sharp rise in power obtained when boost is introduced at approximately half throttle. This limits the intermediate power range of the locomotive since proportions of power between the boosted and unboosted condition cannot be achieved. This difficulty could be almost entirely overcome by introducing boost somewhat earlier and progressively.

This low angle shot accentuates the engine bulk, visually not helped by its restricted length of 50 feet. 10100 heads the 12.05pm Derby to St. Pancras on 20th March 1952 and is seen at Loughborough Midland.

(3) FLUID COUPLING SLIP RANGE

Arising from the tests, it became apparent that serious heating of the oil in the fluid couplings could occur when they are engaged for any appreciable time with engine speeds below approximately 600 r.p.m.

In order to avoid this trouble the four main engine rev counters have been marked at 600 r.p.m. and the drivers instructed not to dwell at engine speeds below that value.

(4) MAIN ENGINE COOLING WATER TEMPERATURE

Throughout the tests, it was noted other than when working in regulator position 2, which gave maximum power output without boost, the cooling water temperatures approached boiling point. This was due to the water circulating pumps being driven by the auxiliary engines which only run at idling speed when the main engines are not boosted. This difficulty could be overcome by driving the pumps from the main engines or alternatively by earlier introduction of the boost as suggested under (2). This latter step would give higher auxiliary engine r.p.m. and therefore increased coolant flow at this regulator setting. In the early stages of the tests and prior to the adjustments to the fuel system, some overheating was experienced with D engine when running boosted. Following the adjustments, all engines were cooled satisfactorily when working in the boosted condition, i.e. regulator position 3 to full open. For example, the maximum cooling water outlet temperature was 175°F during the tests with the 390 ton train load, when the locomotive was worked at high power output for long periods.

(5) MAXIMUM SPEED

At present, the maximum speed of the locomotive under power is 72 m.p.h. This could be increased by suitably altering the transfer gear ratio from the A & B primary to the secondary differential. This would, of course, reduce the tractive effort now obtained between 30 and 60 m.p.h. and rotational speed of wheels and coupling rods may be an overriding limitation.

Originally classified 5P5F (the same classification as the LMS twins), it was later increased to 6P5F. No. 10100 is seen here passing Stanier 8F No. 48696 at Kettering.

Perhaps not surprisingly for a prototype, much time was spent in the works. 10100 is seen at Derby Works in the spring of 1954.

On 16th October 1958 at Manchester Central the engine's train heat boiler caught fire. The loco was taken to Derby where it was dumped in the shed yard and was withdrawn on 22nd November and after being slowly dismantled was finally scrapped in July 1960 having run around 100,000 miles.

THE STANMORE BRANCH

A BRANCH LINE BUILT BY A HOTEL MILLIONAIRE

Stanmore Village railway station was opened on 18th December 1890 by the Harrow and Stanmore Railway, owned by a hotel millionaire by the name of Frederick Gordon. It was a short branch line running north from Harrow & Wealdstone station. Trains were operated by the London & North Western Railway (L&NWR).

The station was located on the south side of the junction of Gordon Avenue and Old Church Lane (the section north of the junction was originally named Station Road), and was noted for its architectural style as it was designed to resemble a village church, including a short spire. It closed to passenger traffic in 1952.

HISTORY

In 1882 the entrepreneur and hotelier Frederick Gordon purchased Bentley Priory, a large country house near Stanmore. He planned to open it up as a country retreat for wealthy guests. Known as "The Napoleon of the Hotel World", Gordon was a successful international businessman, and had earned his fortune through his international hotel chain.

At the time, Stanmore was a rural location, and Gordon decided to build his own railway line from Harrow in an attempt to attract affluent clientele to his country hotel. He negotiated a contract with the London and North Western Railway to operate the Stanmore line on his behalf.

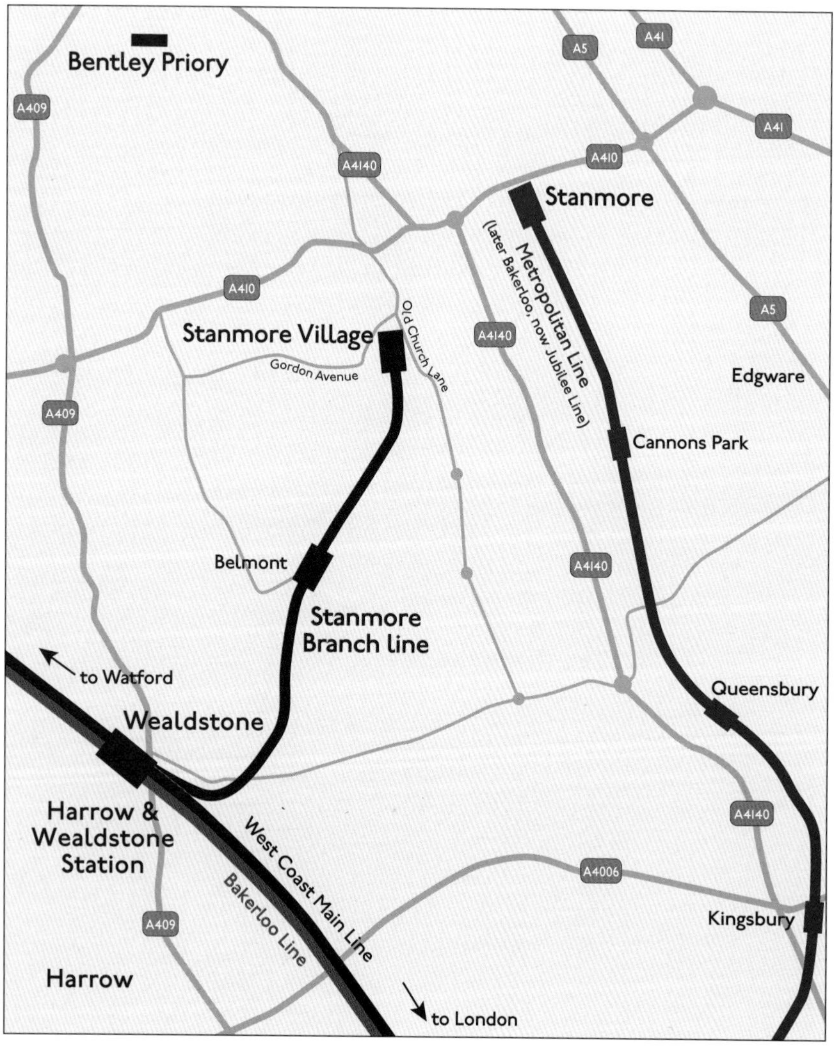

Gordon's scheme met with some local opposition and he was forced to re-route the railway line further east to mitigate objections. The site for a terminus was selected in Old Church Lane in Stanmore. To allay the concerns of the local inhabitants – and to appeal to his well-heeled customers – Gordon commissioned an architect to design an elegant station building that resembled a Gothic-style English country church. The Stanmore branch line opened to great fanfare on 18th December 1890.

The connection to the main line at Harrow & Wealdstone station faced away from London, preventing through trains operating without a reversal; the passenger service was thus operated as a shuttle from Harrow and Wealdstone station to Stanmore, normally run as a push–pull train.

ARCHITECTURE

Stanmore was an affluent and conservative community, and Great Stanmore Parish

Council stipulated that Frederick Gordon's new station building should be of a high-quality design that would blend in with its surroundings. The station was built in a Gothic style, deliberately designed to resemble a small English church complete with a square tower topped with a spire and decorated with gargoyles, a large clock on top of a buttressed Gothic portico, and an ecclesiastical-style entrance door. The station had a single platform covered by a cast-iron and glass canopy. The supporting pillars bore the coat of arms of the Gordon family; three boars' heads surrounded by thistles and roses.

Exterior view of Stanmore Village station – 11th March 1952

DECLINE

The opening in 1932 by the Metropolitan Railway of its own Stanmore station about 0.6 miles (1 km) to the north-east (later served by the Bakerloo line and now by the Jubilee line) introduced a rapid, direct service to the West End and the City of London. This presented strong competition to the village station, which was now operated by the London, Midland and Scottish Railway (LMS). An intermediate station on the LMS branch line was constructed at Belmont in order to attract more passengers, opening on 12th September 1932.

Stanmore station was renamed Stanmore Village on 25th September 1950 in order to distinguish it from the nearby Underground station. Declining receipts led to the passenger service being withdrawn on 15th September 1952, but a shuttle service continued between Belmont and Harrow. The London Transport 158 bus route provided alternative services. A daily freight train continued using the line beyond Belmont.

CLOSURE

In 1963 the entire Stanmore branch line was marked for closure as part of the Beeching cuts. On 6th July 1964 the goods line from Belmont to Stanmore was shut; by then the run-round loop had been removed so goods wagons were propelled from Harrow and Wealdstone station. The last passenger train on the remaining section ran from Belmont to Harrow on 5th October 1964.

The track was lifted in 1966 and the remaining trackbed was purchased by Harrow Council. Sections of the former line were sold off and built upon, but most of the line was left to grow wild. The Stanmore Village platform buildings were demolished in the 1970s for the construction of a road of new houses, September Way, which was built along part of the track alignment. Despite its architectural merit, Stanmore Village station was allowed to fall into ruin. Attempts were made to preserve the building, but it suffered from neglect and vandalism. In 1969 it was redeveloped by a property developer, who removed most of the Gothic architectural features and converted it into a residential property, which still stands today on Gordon Avenue. A plaque mounted on the wall of the house indicates the site of the station.

A view of the station on its last day of operation – 13th September 1952

A wonderful photograph of the ornate Stanmore station with plenty of interest created by the building itself and the period signage to the right of the locomotive, including an LMS notice board. Stanier Class 2P 0-4-4T No. 6409 waits to depart with a shuttle service to Harrow & Wealdstone. It was one of ten engines built from December 1932 to January 1933, their main duties being light passenger work. Note the stovepipe chimney, which were fitted to the whole class, these were eventually changed to the more conventional Stanier design. Although undated the photograph would have been taken in 1946 or before as the engine was to be renumbered 1909 at the end of that year. PHOTO: NEVILLE STEAD © THE TRANSPORT TREASURY

BOVRIL
SANDWICH
puts BEEF into you
STANMORE
6409

OPERATIONAL PROBLEMS OF THE LMS 1939–1945

TIMETABLES, CIVILIAN TRAINS AND THE RETURN OF PEACE

THE PRE-WAR TRAIN SITUATION

In the summer of 1931 no passenger trains were scheduled at a booked average speed of 60 m.p.h. and over, but in the summer of 1939 there were 67 such trains that covered 6,882 miles daily. The acceleration of passenger trains can be seen by the number of trains affected, and the total amount of acceleration. In the eight years from 1931 to 1939 alterations were made to the timings of passenger trains in 16,267 instances, thus saving 886 hours daily.

A review was made to obtain accelerated point-to-point timings, where possible, of all freight trains, so as to secure the best possible performances from the modern motive power stock in relation to the traffic to be moved and to maximise track usage/track occupation. In the seven years 1932 to 1939 accelerations to freight trains numbered 691, saving 384 hours. At the end of 1931 there were only 173 freight trains composed wholly or partially of wagons fitted with the vacuum brake, whereas at the end of 1938 the number of such trains had risen to 339, an increase of 96%, with a total daily reduction in journey time of 150 hours.

THE EFFECT OF AIR RAIDS ON TIMETABLED TRAINS

When war commenced the timed speeds of freight trains was not altered except that certain of the 'fitted' (vacuum braked) and partially fitted trains were reduced in classification. Express passenger trains that, during pre-war years had been accelerated, and ran at high speeds, were decelerated at the outbreak of the war, and at the time of air raids were booked to run at speeds not exceeding 60 m.p.h. at any point. Even with this reduction in booked speed of trains, the dislocation of the pre-arranged plan that occurred under air raid warning conditions when the trains had to be brought to a stand, the drivers warned and the speed reduced to 15 m.p.h. in the case of passenger trains and 10 m.p.h. in the case of freight trains, and subsequently stopped again for the warning to be cancelled, will readily be appreciated. Additionally, passenger trains had to make a stop at the next station after receipt of a warning to enable the passengers to be advised so that those who desired to do so could alight. When the night bombing started in earnest in September 1940, air raid warnings occurred mostly during the blackout hours and extended for many hours, frequently continuously throughout the night, and were having such a strangulating effect on the railways that it became necessary to take some steps to ameliorate the onerous conditions imposed on the speed of trains. With the authority of the Ministry of Transport the speed during raids, commencing on 11th November 1940, was revised to allow passenger and fitted trains to run at 25 m.p.h. during non-blackout hours, 15 m.p.h. during blackout hours, and all other trains at 15 m.p.h. under both conditions. The extent to which train operation was affected will be gathered from the map opposite which portrays the areas that were under 'Red' warnings and the approximate length of time that warnings were in force. During the night of 12th/13th December 1940, it will be observed that practically the whole of the LMS as far north as Lancaster was working at greatly reduced speed for prolonged periods, with all 'Exempted' lighting extinguished.

To give an example, the 19.20 Euston to Perth was subject to a 15 m.p.h. limit from the start, and the loss of time on this basis compared with the booked timing was sufficient to account for a delay of 12 hours in reaching Lancaster. In practice the train would not sustain delay to that extent as its progress would be so slow that the raid would have terminated before the train had reached Lancaster. It would, however, be some seven or eight hours late at the end of the raid even if the running was not interfered with by damage to the railway. Similar conditions applied to all other trains, both passenger and freight, proceeding north, south, east and west, and so planned working was entirely destroyed. It is not surprising that in the light of experience a further relaxation in the speed restriction was made on

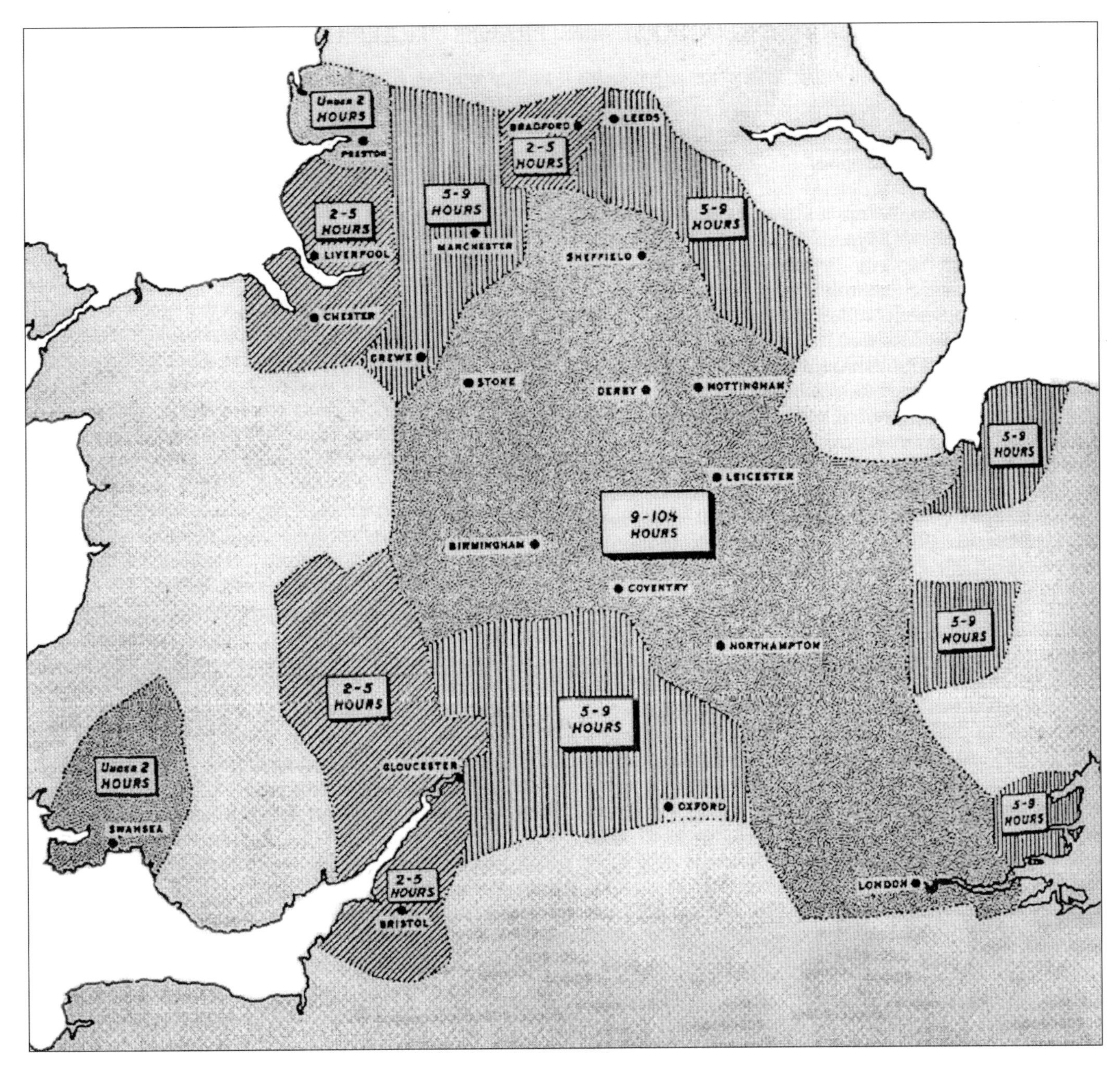

17th February 1941 when all classes of train were permitted to run at normal speeds in non-blackout hours (this removed the necessity to stop trains to warn drivers and again to withdraw the warning during daylight). The speed restriction was increased to 30 m.p.h. during blackout hours. The stop to advise passengers was also abandoned as in practice only a few passengers availed themselves of the opportunity to alight. The air raids followed one another with such frequency and were of such long duration that recovery from one could not be made before another attack commenced. Thus the effect was cumulative and the Operating Officers were faced with a constant struggle to overcome the seemingly almost impossible task of moving passengers, goods and coal, essential to the war effort and the life of the nation.

RAIDS ON TRAINS AND THEIR EFFECTS

The following facts and figures convey the extent to which railway working was held up, apart from the physical damage sustained.

During the darkest months of the year – October 1940 to March 1941 – the average rate of movement over LMS lines in England and Wales fell by 5% for passenger trains and 20% for freight trains, and by 14% for both classes together, compared with the same months the previous winter, when the railway was already working under war conditions. Similar figures for the months of October, November and December 1940 were even worse, being 8% for passenger and 27% for freight, and 19% for all trains. To that extent each train on average took longer to pass over the line. The effect on traffic in and around the London region is demonstrated by the figures for wagon miles worked in the Willesden and Kentish Town control areas that included the LMS lines approaching London from the north and in London over which traffic to and from London and via London to and from the south-east and southern counties normally passes.

It is reasonable to assume that, but for the blitz, the traffic flowing via London would at least have been equal to that passing during the corresponding period of the previous year, yet it was reduced by 43% in the Willesden area and 34% in the Kentish Town area.

	WAGON MILES	
	Willesden Control Area	Kentish Town Control Area
11 weeks ended 25th September 1939	13,597,355	9,126,672
11 weeks ended 25th November 1939	7,736,192	5,993,174
Decrease – number	5,861,263	3,133,498
Decrease – %	43	34

A further example in the Liverpool District, where in the final two weeks of May 1941, when Liverpool was suffering from the effect of seven consecutive night raids at the beginning of the month, the movement of freight decreased by 47%, compared with the corresponding weeks in the previous year. But the goods got through, if not by one route, then by another. And all the traffic proffered for conveyance by the railway was carried. The extent to which shunting operations were retarded is exemplified by the results taken in a large traffic yard on the night shift under three conditions.

		NUMBER OF WAGONS SHUNTED	
A	No 'Red' warning	502	No physical damage in any instance
B	'Red' warning in operation the whole time	344	
C	'Red' warning the whole time. Anti-aircraft guns very active	140	
The loss under condition B was 31%, under C 70%			

Heavy demands were made on the Company's resources of locomotives owing to slower running, it being necessary to turn out additional locomotives to complete jobs and to take up return workings, the engines for which had not arrived. This created shortages, and engines of the correct types were not always available for trains. The servicing schedules could not be maintained. During the height of the blitz each locomotive was in use between 7% and 9% longer than at the corresponding time of the previous year. The raids caused the most extravagant wastage of manpower, and a great burden was thrown upon the staff, particularly train crew, who were called upon to work exceptionally long hours. Owing to the slow progress over the line, a train that would normally occupy a set of train crew only a portion of the day frequently required three sets of men to man it before it reached its destination, at a time when manpower was at a premium. Such requirements created shortages that reacted on the number of hours worked, the effect of which is seen in the average weekly figures of the number of cases of engine drivers on responsible duty over 10 hours in comparable periods of 1939 and at the height of the blitz in 1940, that rose from 2,997 a week to 15,780, whilst engine driver turns of responsible duty exceeding 14 hours in length increased from 16 to 2,341. A similar state of affairs existed with regard to firemen, and the figures for guards rose from 2,230 per week to 12,096 (over 10 hours) and from 16 to 2,018 (over 14 hours). Train crew, shunters and goods shed workers were also required to take duty on practically every Sunday to keep pace with the work. It is a well-known fact that drivers, firemen and guards are called out from

Prior to The Blitz glass was removed as an ARP precaution from many London stations, as here at Euston station on 19th July 1940.

their homes to work special trains, and in emergencies. The conditions prevailing naturally created a greater number of emergencies owing to the appropriate men not arriving to take up their return workings. Frequently men off-duty could not be located during air raids by reason of the fact that they were in air raid shelters. No blame attaches to them in this respect as they were acting in accordance with the public policy. Others were bombed out of their homes, sometimes for a second or even third time. The tonnage of merchandise traffic handled at Goods Sheds in London during the three months of September, October and November 1940, compared with the preceding three months, decreased by nearly 305,000 tons or 38%. Whilst for the most part cartage operations ceased during the hours of darkness, the results of enemy action reduced their efficiency. In London and other cities, considerable additional mileage and time was involved in making detours necessitated by the closing of roads, following air raids, and further time was lost in locating tenants of bombed premises or bringing goods back to the station as undeliverable.

ROAD TRANSPORT AND CARTAGE

During 1943 the LMS road transport department conveyed the highest tonnage ever recorded, being 18,150,107 tons, an increase of 38.05% compared with 1938. As far as cartage was concerned, the 1943 figure was 14,599,515 tons conveyed, an increase of 18.89% compared with 1938. On the other hand, 1943 parcel traffic decreased by 46.48% compared with 1938. At the end of 1943 there were 3,579 LMS motor vehicles, 301 more than pre-war, and 17,009 horse drawn vehicles, that was 1,497 more than the pre-war total. Considerable economy in mileage reduction was made in the London area with the LMS and LNER co-operating in order to save overlapping cartage units. There was also a scheme involving all four main line railways in the London Postal District. Up to 1943 the number of LMS horses dropped by 631 to 5,819, being due to the inability to obtain them. Also, where there was a shortage of carters due to their call-up, this required female carters to be trained, of which there were 229 at the end of 1943.

CIVILIAN TRAFFIC

Early in 1939 the railway companies had to face the task of endeavouring to foresee the effect the war would have to passenger timetables with air raids, and the requirements of the armed forces a major consideration. This resulted in the paring down of ordinary peacetime trains, the possible suspension of dining cars and sleepers and also for the reduction in train speeds as a safety measure in order to keep within the maximum average speed between stations of 45 m.p.h. These war timetables had been completed in July 1939 and on Monday 11th September were brought into force. Generally speaking it worked well, but some adjustments were made and services augmented. All but 28 restaurant cars were withdrawn in May 1942, and even these came off in April 1944. At the same time sleeping car accommodation on certain heavy trains was taken over by the Ministry of War Transport and limited, so that the first call on sleeping berths was given to passengers travelling on urgent Government business. The cessation of Summer Time in the first war winter, with the earlier blackout, made many businesses change their hours of work to enable staff to reach home before dark. This obviously resulted in the evening rush hour starting and finishing an hour earlier, with the result being the morning peak period reduced from two to one and a half hours. Workmen's trains also

A scene familiar to many railway workers after overnight air raids, this being Derby Midland station after it was hit in January 1941, killing passengers and rail employees.

featured in the altered conditions as factories sprang up throughout the country, often away from towns, involving more and more trains. As an example, on the LMS at Chorley, 220,500 journeys were made each week by factory workers alone.

The blackout caused the late running of trains as passengers had difficulty in finding seats when blinds were down and soldiers with kitbags and equipment made movement through corridors slow and difficult with the loading and unloading taking more time in the restricted lighting. Also with railway staff having been 'called up', their replacements were less skilled and less able bodied to handle heavy and bulky packages as their predecessors.

Holiday traffic was a further problem. On 21st December 1939 the amount of trains required on this day alone for servicemen's leave was 73, and the total extra trains to run on the last 13 days of that month for the forces totalled 385, but for civilians during the same period no less than 2,693 were required with the peak being on 23rd December when 644 trains were required. Such effort could not be expected to continue and the services were appealed to for help and the public requested not to travel unless it was absolutely essential, hence the wartime phrase *"Is your journey really necessary?"* At Christmas 1940 it was made clear by the REC that there would be no additional services or cheap travel. The summer of 1941 saw the public requested to take holidays without travelling. Traffic remained heavy and the same appeals were made in 1942 with little or no effect as the months from May to September saw traffic rise to new heights, exceeding the 1938 figures by 74%. The appeals to the public were not therefore very successful as doubtless the strain of war years meant that a reasonable summer holiday was more necessary than ever and folk felt justified in visiting family or taking a holiday. In March 1943 the Ministry of War Production again called for staggered holidays but similar levels as in 1942 still prevailed. To give an example of the disregard to "stay at home" the flow of traffic to Blackpool during the period May to September was: 1936 – 2,426,248; 1942 – 2,068,682; 1943 – 2,598,600.

The following table shows the number of tickets, including season tickets, issued during the war years in comparison with 1938:

Year	Number	Increase/decrease	%
1938	410,911,329	–	–
1939	372,189,765	-38,721,564	-9.42
1940	326,655,468	-84,255,861	-20.5
1941	363,931,496	-46,979,833	-11.43
1942	418,851,453	+7,940,124	+1.93
1943	440,906,167	+29,994,838	+7.30
1944	437,267,369	+26,356,040	+6.41
1945	426,177,730	+15,266,401	+3.72

The numbers of originating passenger journeys do not tell the full story as they do not include through journeys commencing on other railways, e.g. troop movements and leave journeys starting in the southern counties that passed over the LMS railway possibly for distances of 400 miles or more. These figures are the only measure available of the increase in the volume of passenger traffic that show that the peak passenger travel was in 1943. It should also be noted that comparing 1945 to 1938, 3.72% more passengers were carried, with a reduced passenger train mileage of 29.54% that indicates a much heavier loading of trains. The decline in the number of originating passenger journeys in 1945 was due to a marked decrease in the number of workman's journeys and in local travel. Long distance traffic continued at a high level exceeding that in 1944 and pre-war.

The number of passengers who commenced their journey on the LMS increased enormously when comparing the year 1938 with 1945. In 1938 the figure was 32,485,726 and in 1945 it was 199,327,333; that includes monthly return, period excursion, tourist trades, contract, other reduced tickets, and services travel with warrants, both duty and leave. The train miles run in 1944 and 1945 compared with 1938 were:

Year	Miles	Increase/decrease compared with 1938	%
Coaching			
1938	103,668,285		
1944	70,457,214	-33,211,071	-32.04
1945	73,040,435	-30,627,850	-29.54
Freight			
1938	55,577,730		
1944	63,980,878	+8,403,148	+15.12
1945	59,733,877	+4,156,147	+7.48
Total			
1938	159,246,015		
1944	134,438,092	-24,807,923	-15.58
1945	132,774,312	-26,471,703	-16.62

Whilst the 1945 coaching train mileage showed an increase over the previous year being 3.07% greater than in 1944, it was 29.54% less than in 1938. The increase over 1944 was due to relief trains to cater for traffic that could not be dealt with by the booked services, and to additional trains run for Whitsun and August Bank Holiday as far as stock, available staff and essential traffic permitted. The decrease in freight train mileage in 1945 was due to the sharp decline in traffic following Victory in Europe, whilst after April coal traffic was lower than in 1944 and mineral traffic less than any time during the war years.

Apart from the Easter holidays, when the Ministry of War Transport placed limits on the number of long distance trains that could be run, no such directions were issued for the remaining bank holiday periods and so the LMS ran as many trains as possible in line with the available staff. Needless to say all trains were loaded to capacity and there were instances of passengers not being accommodated at originating stations and joining points. Various interesting Government directives were issued, one of which was in October 1941 which stated that all First Class accommodation be withdrawn from trains commencing and terminating in the London Passenger Transport area. The situation became farcical, as, after the order came into use, there was a rush for the First Class seats that resulted in people standing in them whilst other parts of the train were fairly empty. Referred to earlier was the curtailment of sleeping coaches, and a further attempt was made in 1941 to increase capacity, when arm rests in all Third Class compartments of modern design were screwed back. The situation was summed up by the Chairman of the Company, Lord Royden, remarking at the 1944 Annual General Meeting that: *"Railway travel now places a considerable strain on one's physical powers of endurance, not to speak of one's power of resistance to irritation"*. As in the First World War petty theft increased with the LMS losing 400,000 hand towels in 1941 alone, what was worse was the slashing of upholstery and the breakage of windows and fittings that reached disturbing proportions. Also of interest was that limitations had been placed on the number of times that parents could visit their evacuated children. The Ministry, in 1945, felt that this was no longer justified provided they did not visit more than once per month. In November 1945, the rationing scheme for the concession travel for the wives of servicemen was cancelled for journeys over 30 miles. On 12th October the LMS President intimated that he had agreed informally with the Minister of War Transport to additional trains, that were the strengthening of regular services, could be run for particular parties, or where the additional traffic arising in connection with special events could not be dealt with on regular services provided that any such trains were available to the general public and no special accommodation was reserved. Cheap day tickets, excursion and other reduced fare facilities withdrawn in the early part of the war were not reinstated and this continued to have a limiting effect on passenger travel, although during the summer months trains were seriously overcrowded with people intent on enjoying a holiday, in many cases for the first time in six years, and with forces personnel on ordinary and demobilisation leave. The 1945 Christmas traffic was exceptional with the number of passengers for the three weeks ending on 7th January 1945 totalling 2,665,000. Cases occurred when passengers could not be handled on the trains for which they had presented themselves, that were however cleared by subsequent services or the running of additional trains. In all 707 additional passenger trains were run between 20th and 27th December.

Such was the effort required by the staff that many showed the strain, whilst sickness amongst train crews and yard staff was heavy and in addition, a number of men simply failed to turn up for their duty.

During the five days from 27th to 31st December 1944, absent staff totalled 6,515 or 13.1% of the total staff at the depots concerned situated in the Western, Midland and Central divisions.

SPORTING EVENTS

The programme for the 1944 Flat Racing Season approved by the Government was similar in regard to venues and the number of days in 1943. Whilst the railway companies had authority to convey racehorses by rail in connection with certain 'Open' events at Newmarket in 1943, the Ministry of War Transport (MWT) imposed a total embargo on the acceptance of horses for transport during the 1944 season. In both 1943 and 1944 no special trains were allowed for racegoers to or from meetings. The MWT approved a limited resumption of steeple-chasing for the 1944/45 season subject to there being no rail movement of horses or spectators. Rail facilities were granted by the MWT for racing pigeons in 1943 and 1944 with 30 special trains run by the LMS.

MAIL TRAINS

The LMS had an agreement with the General Post Office dated 10th December 1930 for the conveyance of mails and a number of trains were specified to run at certain times on specified days of the week. Some trains were able to throw out and collect mail at speed but the prohibition of external lighting caused this practice to be discontinued early in the war that was not reintroduced until 1st October 1945.

ROYAL TRAINS AND SPECIAL TRAINS FOR IMPORTANT PERSONS

The LMS Royal Train was stored at Wolverton and used for tours by their Majesties the King and Queen, and up to the end of the war the LMS ran 151 tours on the Company's system, covering 35,690 miles. Tours could last for several days while they visited Army, Navy and Air Force bases. Care was taken as to where the train was stabled, and where possible near a tunnel to enable the train to be drawn into it should an air raid occur. The LMS train was also loaned to other companies – LNER, 55 tours; GWR, 15 tours; and SR, seven tours. The Prime Minister, The Rt. Hon. Winston Churchill, also had a special train known as the 'Rugged' train, comprising of a 1st class brake, two saloons, a 1st class vestibule, 1st class dining car and sleeping cars with one saloon specially adapted for the Prime Minister's use. This train was first used in September 1941 with a total of 36 journeys made during the war period that totalled 16,000 miles. In September 1945 the Rt. Hon. C. R. Attlee expressed a wish that the special coaches forming the 'Rugged' not suitable for general purposes should be kept for his disposal, and arrangements were made to hold the corridor brake first, saloons 803 and 804 and the vestibule first, the remainder to return to general service. A train known as the 'Alive' train for the Commander-in-Chief, American Forces in Europe was supplied by the GWR. This train did 41,022 miles on the LMS rails and was shipped across to Europe on 14th December 1944. Finally a further train known as the 'Rapier' train was formed for the Commander-in-Chief, Home Forces. This train was provided by the LNER and ran until 11th July 1945 by which time it had travelled 26,689 miles on LMS metals.

1945 – PEACE RETURNS

The cessation of hostilities brought no let up for the LMS with thousands still travelling with hundreds of special trains. Combat troops were returning on leave being replaced by reinforcement personnel. Prisoners of war and internees were returned home, Soviet Nationals were repatriated, Belgian, Dutch, Polish and Canadian servicemen were also returned home and all the time servicemen were arriving from the Far East by ship. American servicemen were returning to the USA at the rate of 15,000 at a time on the Liners *Queen Mary* and *Queen Elizabeth*. From VE Day to August these departed from the Clyde, requiring 37 special trains each time from the Salisbury area.

The situation was eased when the ports of Tilbury and Southampton could again be used thus taking the pressure off the Mersey and the Clyde that had been the only ports used for the previous five years, both served by the LMS. As far as demobilisation was concerned, that commenced on 18th June 1945, the army had nine centres for troops serving at home. For army overseas personnel a place was allocated to each arrival port as follows – Carlisle (from the Clyde), Strensall near York (from the Mersey), Shorncliffe (from Folkestone), Reading and Oxford (from Southampton and the Bristol Channel). The RAF had its own plan with demobilisation centres at Hednesford (Staffordshire), Cardington (Bedfordshire) and Kirkham (Lancashire). The Navy had given no indication as to its plans and demobilisation was carried out direct from the various Naval barracks. Special trains were planned, but with few using them they were withdrawn, with block reservations on ordinary trains being utilised. Extreme pressure was experienced at some of the demobilisation centres and so others were opened such as at Aldershot and Woking. A centre was opened at Prestatyn for Pioneer Corps personnel. Following documentation these men were transferred to Stalybridge where they were passed out of the army with a special train being run each day from 4th September to 1st October.

From 18th June to 31st December 1945, 410 special trains were run in connection with demobilisation. Obtaining the exact arrival time of ships was often a problem, resulting in valuable empty stock trains awaiting their arrival. It was a common occurrence for ships to arrive earlier or later than planned, an example being when the *S.S. Georgic* arrived at Liverpool with 5,000 personnel. Considerable difficulty was experienced in obtaining the ship's time of arrival from the Captain. This matter was taken up with the Ministry of War Transport with a view to obtaining definite arrival times of ships.

There were thousands of prisoners of war in the country that required special trains for their movement, an example being 14 trains to move 12,400 to Northern Ireland in early 1945. At the same time 2,500 prisoners of war were moved to the Isle of Man requiring five special trains, some of which later returned to help with the 1945 harvest. With POWs arriving at Southampton from the continent in large numbers, frustration was experienced with what was expected from the railways. As an example prisoners were being loaded into two trains at Southampton before advice was received by the LMS that the destination was Forres in the north of Scotland, a journey of 585 miles. In the event engines and men were provided at London junctions within six hours, but the War Office was notified that such long distance hauls could not be undertaken in future unless ample notice was given. The number of POWs increased to such an extent it was necessary for a redistribution in March 1945 when 29 special trains were run in a three day period. Prisoners were also carried on ordinary passenger trains with instructions issued by the War Office as below following complaints from the public that they were standing whilst POWs were comfortably seated.

- No reserved accommodation to be provided for parties of less than 20.
- Parties for whom no reservations were made must not occupy seats if ordinary passengers are standing.
- Parties of 20 or more to be given reserved accommodation at the rate of 8 to a corridor compartment and 12 to a non-corridor compartment.
- Ordinary passengers must not be removed from a compartment to make room for a POW.

A camp in the Windermere District was occupied by high-ranking German Generals from where parties of these individuals travelled to London by ordinary train for interrogation. In October evidence was being gathered for the Nazi Nuremberg War trials when a party of 42 travelled from London to Windermere in First Class accommodation at the request of the War Office.

It goes without saying that members of the allied forces were often in a sorry plight once the camps they occupied were overrun by the allies and ambulance trains were often required at urgent notice to get them to treatment centres or hospitals. A total of 83 special trains conveying 26,592 ex-prisoners of war were run between 10th April and 1st June 1945 with numerous small parties being conveyed by ordinary passenger train. It was not until 12th October that the first ex-prisoners of war returned from the Far East and from that date until 10th November, 26 special trains were run from the Mersey.

Christmas leave for the forces in 1945 provided problems as little restriction was imposed by the Service Departments. WD Depots and Establishments closed for the holiday with 335 special trains worked over the LMS between 20th and 28th December. In some cases special trains that had been programmed were either lightly loaded or cancelled, as the number of personnel did not reach expectations. To quote a couple of instances – eight trains were booked to convey RAF personnel from Kirkham to Glasgow, York, Manchester, Birmingham, Cardiff and London on the morning of Sunday 23rd December. Two of these trains were cancelled and the remaining six conveyed on 25-50% of the numbers anticipated. It later transpired that the men had been released at noon the previous day and consequently 4,000 personnel had made their own way home thus contributing to the overcrowding. A further instance concerning the conveyance of US servicemen from London to Lytham found that only 120 passengers were on the train instead of the 500 expected, the train was terminated at Preston.

A scheme was devised by the Ministry of Health for parties of Dutch children to come to this country for recuperation lasting two to three months. The first batch of 500 accompanied by 100 adults arrived at Tilbury on 11th February, with a special train provided to take them to Coventry. Other groups followed and the first party returned to Holland in May.

The Liberation of Europe also brought problems to the LMS in that stores already en route had to be cancelled; on the other hand the liberation of the Channel Islands on 10th May required large quantities of food and civilian supplies to be sent with all possible speed. Liberation also led to the disposing of large quantities of ammunition in the sea off the Mull of Galloway. Almost a train a day was sent from various WD depots to ports on the north coast such as Cairnryan, Silloth and Workington from the middle of June to the end of December. A quantity was also sent to the south coast for dumping in the sea. A total of 270 special trains were run by the LMS for this purpose. Eighteen special trains ran from the USA depot at Kimbolton carrying bombs in 720 wagons to south Wales for shipment to the Pacific front and following this, a further 15 special trains were run to Swansea for shipment back to the USA.

With many Ordnance Depots being closed at the end of the war it was necessary to transfer stocks and supplies to those depots of a more permanent nature that required special trains as the following examples show: 13 special trains from Lytham (USA depot), to Manchester and Southampton for shipping back to America; 55 special trains of petrol and oil were run from Grangemouth, Birkenhead District and Carnforth to various other depots; 14 trains of armoured vehicles from Salisbury to Redditch; 122 trains from Ashchurch to Southampton, South Wales and Birkenhead returning stores and supplies back to the USA. Twenty trains of armoured vehicles from Winchester to Leicester and Gateshead; 35 trains of heavy tanks from Rainford, near Wigan to Histon, near Cambridge, and 34 trains of armoured vehicles from Coventry and Stechford to Hereford.

There were other more peaceful requirements the LMS had to deal with such as the resumption of banana traffic, the first consignment of which arrived at Avonmouth on 31st December 1945. This traffic used special vans fitted with steam heating to ripen the bananas en route to their destination. The problem was that all these vehicles had been used during the war for the transport of meat and other commodities and required the steam heating apparatus to be restored with 1,500 vehicles completed in time for the first consignment. New housing became a Government priority as the war neared its end with prefabricated houses being imported from the USA. Each house was shipped in eight packages weighing 7.3 tons and 30,000 were programmed of which 8,600 had been received up to the end of November 1945.

A great many other instances could be given detailing the efforts made by the LMS and its staff rose to every occasion in the movement of just about everything.

CARS AND SLEEPING COACHES FOR STAFF

Due to the shortage of footplate staff and guards in 1944, provision was made at Bletchley, Northampton and Nuneaton to accommodate staff transferred from other centres. With lodgings difficult to find, camping coaches, sleeping and kitchen cars were utilised to provide sleeping accommodation and meals. Games, wireless, magazines and darts were also provided for which the men concerned paid 28 shillings (£1.40) per week. Similar arrangements were made at Rugby and Cricklewood in the early months of 1945. Facilities for lodging were also provided at Gloucester for men booking off away from home due to the difficulty of the men obtaining private lodgings. The recruitment of staff from Ireland and Eire caused an accommodation problem that was solved with the temporary use of a British Sugar Corporation Hostel at Nottingham.

OPERATING STAFF

The total number of staff employed in the Operating Department in all grades, clerical, supervisory and wages including shed staff in 1938 was 118,367 and at the four weeks ending on 31st December 1945 it was Goods and Traffic, 88,801 and Motive Power, 47,696 making a grand total of 136,497. Such was the shortage of staff, that commencing in February 1944, the loan of troops from Royal Engineers, Railway Operating Companies was utilised to ease the situation. The LMS share was 351, being Firemen, Shunters and Shed Staff, all of which were withdrawn on Monday 28th August 1944. At the end of 1945 the number of Operating Staff still serving with the forces and who were expected to resume service with the LMS was 19,808. The number of staff already demobilised and had resumed duty by that date was 1,380 whilst casualties and resignations totalled 1,225. During the same period the number of women employed in male wages positions was 12,776 or 9.1% of the total wages staff, at the end of December 1944 it was 15,262 or 11.5% of the wages staff. At the end of 1944 the number of women employed as signalmen, porter signalmen, passenger guards and porter guards was 1,006.

THE MEASURE OF SUCCESS ACHIEVED

Figures have been quoted to demonstrate the effect of the blitz on railway operation, but what may be even more surprising is a measure of the extent of the success that attended the efforts to circumvent and surmount the unprecedented handicaps. In the nine months from September 1938 to May 1939 inclusive, under peacetime conditions when there was no blackout, the number of passengers in the long distance category was 20,231,064, whereas in the corresponding months of 1940 and 1941 – the period of the blitz – the total was 29,125,124, the increase amounting to the remarkable figure of 8,894,060 or 44%. The demand for short distance travel was consistently less than in peacetime, so that comparisons are not relevant. On the freight side, the loaded wagon miles – that are the truest means of effective operating work performed – increased in the same period by 207,440,750 wagon miles, or 21%. The quantity of freight handled in goods sheds rose by 2,876,000 tons (17%) and the volume of freight carted by LMS vehicles by 763,700 tons. Assuredly, the Chief Operating Manager, Mr. T. W. Royle, together with his operating staff, accomplished superhuman feats in the face of the greatest upheaval of railway working of all time. It should be mentioned that the Operating Manager, T. W. Royle was appointed a vice-President of the LMS on 1st September 1944 when S. H. Fisher succeeded him and who wrote the war report for the final wartime years.

STOCKPORT EDGELEY

An early morning scene at Stockport Edgeley shed (9B) with locomotives prepared and ready for their day's duties. From left to right we see Stanier 'Black Five' 4-6-0 No. 44946, Standard 7MT 4-6-2 No. 70015 APOLLO and two more 'Black Fives', Nos. 45225 and 44826. Although the image is undated it is likely to be sometime between 1965 and 1967 when both No. 70015 and 45225 were allocated to 9B.

Stockport shed was opened for use on 24th May 1883 by the London & North Western Railway. After the grouping in 1923 the shed was placed in the Western 'A' Division as a sub-shed of Longsight (16). The LMS replaced the shed roof with a 'single pitch' design in the early 1930s, and allocated the code 9B in 1935. The 60 foot turntable, which dated from 1927, was fitted with a vacuum tractor at a cost of £210 in the busy wartime conditions of 1942, at which time a new water column and ash pit were also added. The shed became home to WDs, Stanier 5MTs and 8Fs as well as BR Standard locos from the late 1950s, replacing older engines as they were withdrawn. Stockport Edgeley closed on 6th May 1968.

"the usual charge between companies for the use of a turntable was 6d per movement." This might possibly be taken to imply that a turntable was already in existence at Bath and that payments were being made for its use. If an original turntable had been in existence since 1876 it would appear odd that it was considered necessary to replace it

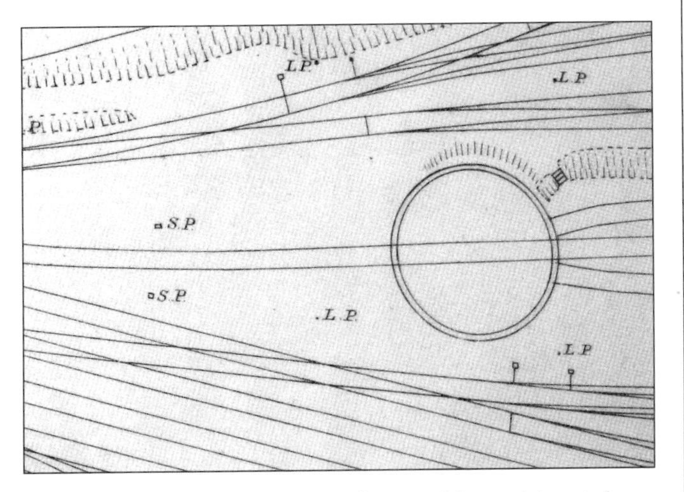

An early close up of the original turntable position, taken from maps of 1883/1885, showing three roads leading off the turntable rather than the one road provided.

just two years later in 1878. However, as Mike Arlett has suggested, this could be because it was in the way of a proposed enlargement of the S&D shed to four roads. Unless further evidence comes to light we can only say that the possibility that a turntable existed at Bath prior to the installation of that of 1878 should not be discounted.

The original position of the proposed replacement turntable is shown on the sketch below from which it will be seen that the turntable was intended to be located on the north-west of the site, providing nine roads in a partial roundhouse fashion linked to the access to the S&D shed. Information is taken from the 1931 LMS drawing showing the proposed location of the 60 foot replacement turntable in that part of the site close to Midland Bridge Road.

The installation of the replacement turntable of increased size at Bath in 1935 featured a Cowans Sheldon 60 foot model. This firm, who were based in Carlisle, specialised in cranes but from the mid-1850s manufactured locomotive turntables producing the first balanced turntables for early railways. They then progressed to manufacturing non balanced or articulated turntables having supplied large numbers to British railways and to overseas railway companies. The advantage with articulated tables is that the load on the centre pivot is virtually halved and necessitates a much shallower pit; they also have advantages with ease of operation and simplicity of maintenance. Turntables were manufactured to suit standard, broad and narrow gauge lines with standard sizes being 60, 65, 70, 80 and even 100 foot. They could be powered manually, electrically or by vacuum power from the locomotive being turned, the firm having invented their own patented vacuum turning system. Sheldon

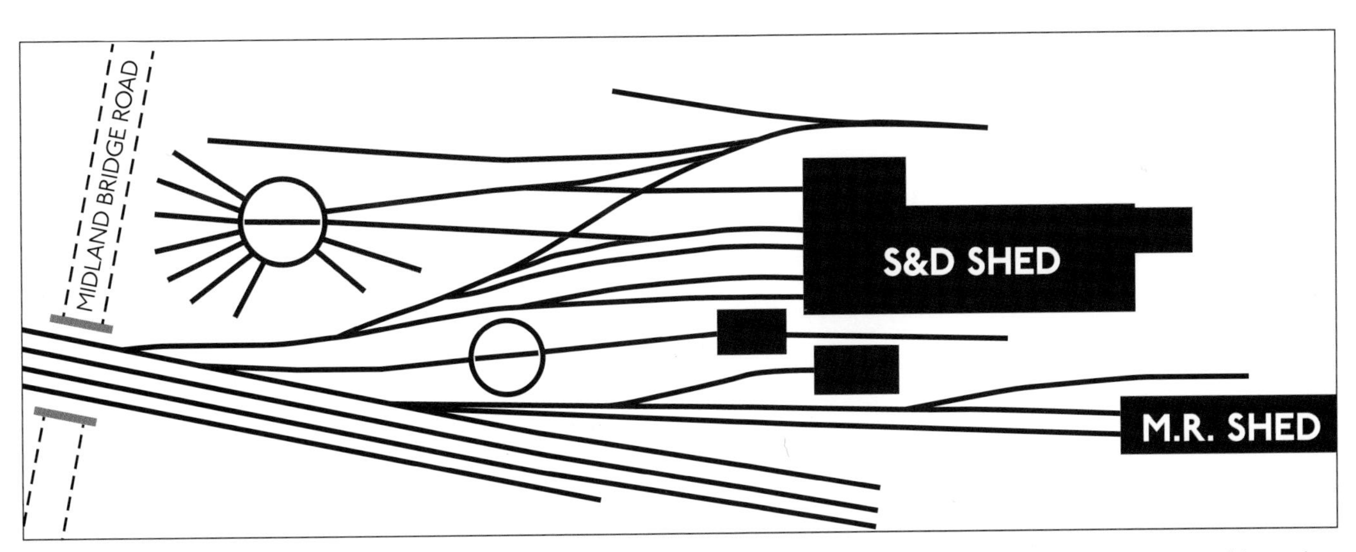

Information taken from an LMS drawing of 1931 showing the proposed location of the 60 foot replacement turntable in the north-west corner of the site.

Another view of the turntable in action is provided by the presence of a stranger in the shape of U Class 2-6-0 No. 31639 which worked in with 34015 EXMOUTH on the RCTS special of 2nd January 1966, the original closure date for the line. PHOTO: DEREK FEAR

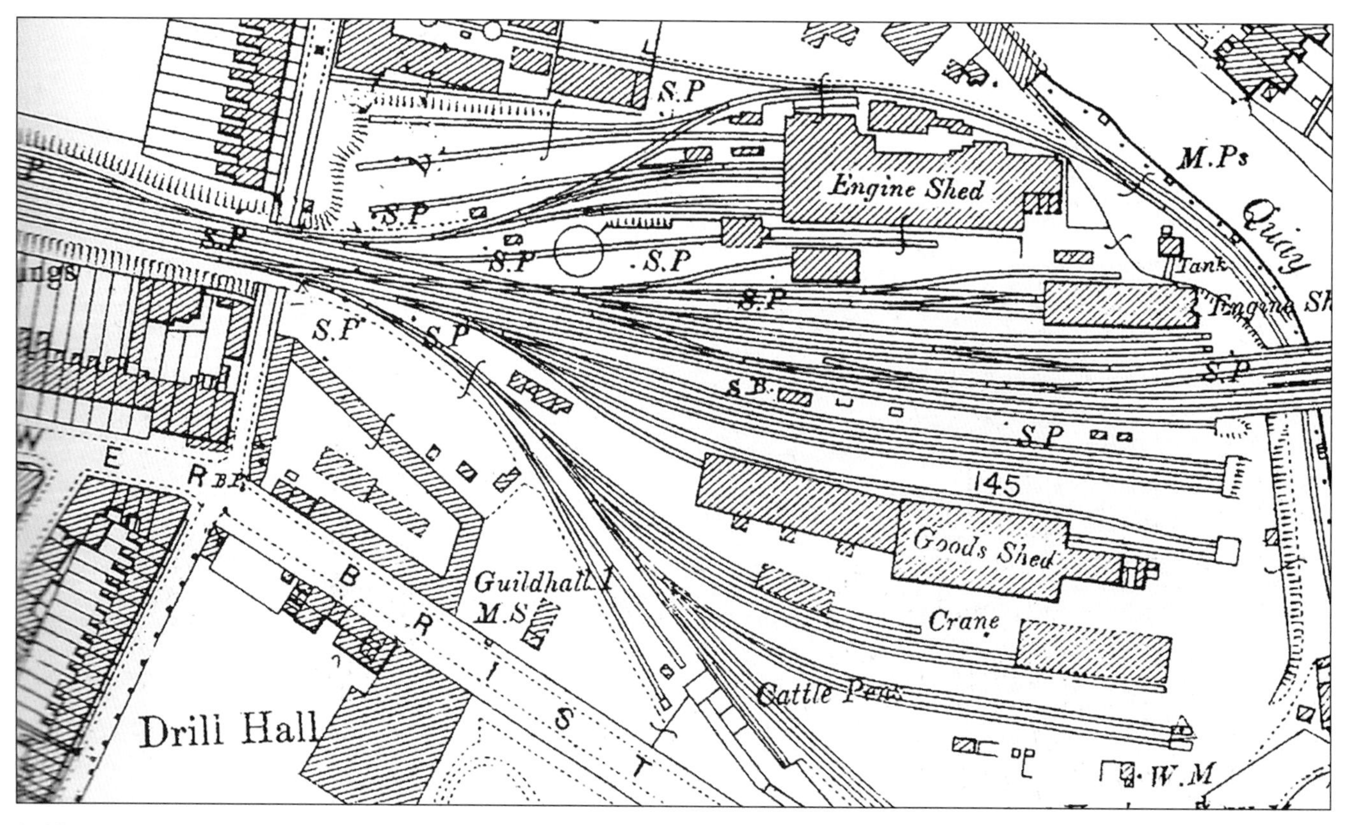

A 1904 map shows the position of what was possibly the original turntable, with just one road leading from it, located to the left side of the Midland shed which was constructed following the opening of the line from Mangotsfield.

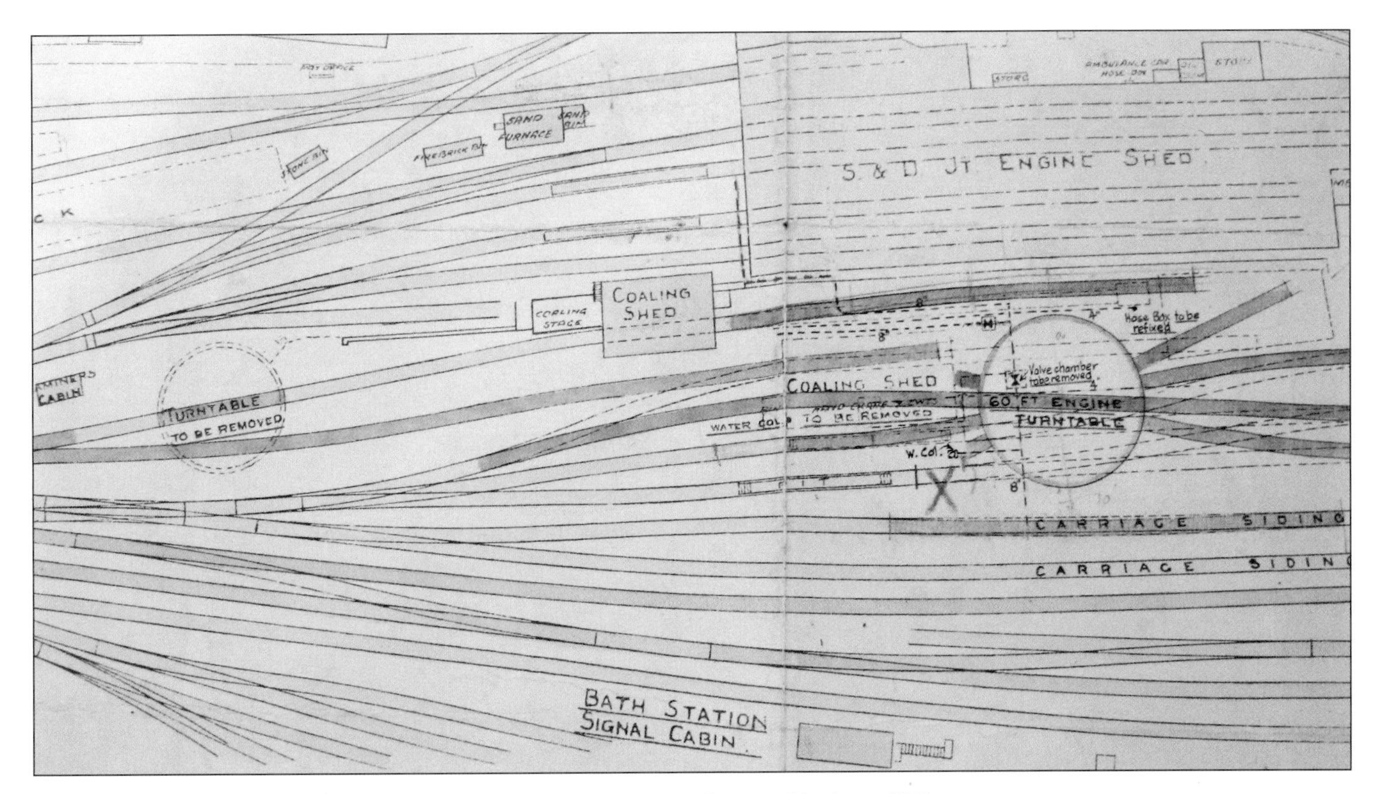

LM&SR Bath Engine Sheds. Diagram showing revised position of turntable from 1935.

Cowans continued in business after mergers and takeovers into the 1980s until manufacturing of cranes ceased in 1987.

In the event a revised position for the new turntable at Bath was chosen in front of the Midland shed, all locomotives having to use the turntable to gain access to the shed. Quite why this position was decided upon is not clear especially given the general rule that sole access to a shed across a turntable was generally avoided as its failure could trap the locomotives housed inside.

However, the MR shed was, by the 1930s, mainly used for repairs and the S&D shed was designated as the running shed so no doubt this factor did not weigh unduly with the authorities in the decision to site the turntable where they did. It lasted here until closure of the line in 1966. On occasions when the turntable was out of action recourse was had to sending locomotives, sometimes up to four or five at a time coupled together, to turn on the triangle at Mangotsfield. This happened in April 1953 just over two years after renovation of the turntable, when it had to be taken out of service for a week following mechanical trouble. This trip of some 20 miles to Mangotsfield and return was catered for by the provision of special paths in the working timetable, allowing about an hour for the return journey. This procedure was captured by Ivo Peters in Volume 1 of his book *Somerset & Dorset in the Fifties* where a series of three photographs illustrate 44146, 43939, 34042, 53805 and 44848 coupled together undergoing these manoeuvres. It is somewhat surprising that the line had to wait until 1935 for a replacement turntable of sufficient size as the famous 7Fs, which had been introduced in 1914, could not be turned on the turntable then existing at Bath and it was to be 20 years before they could do so.

One of the operational problems of having the turntable in its revised position was that, when larger locomotives such as 9Fs and Bulleid Pacifics were drafted in to work on the line, there was a danger that, if these types were positioned a little far forward, the leading buffers would, as the turntable revolved, foul wagons standing on the adjacent coal stage road. Just how tight the clearances were is illustrated by the fact that the 9Fs had a wheelbase of 55 ft. 11 ins. and an overall length of 66 ft. 2 ins., whilst the Bulleids had a wheelbase of 57 ft. 6 ins. with an overall length of 67 ft. 4¾ ins. In fact before the SR authorities allowed a Bulleid Pacific up the line for the first time in

March 1951 they checked with Bath shed that they had a 60ft turntable available. The proximity of wagons on the coal road could of course have its advantages for, by some judicious positioning of the tender or coal bunker on tank locomotives, it was relatively easy to filch some coal from an adjacent wagon to top up supplies!

For the record, the other turntables on the S&D were located at Evercreech Junction (56 ft.), Highbridge (49 ft. 9 ins.) and Templecombe (50 ft.). There was also a turntable at Wimborne (44 ft. 9 ins.) on the old route used before the opening of the Corfe Mullen cut-off. Turning at the Bournemouth end was accomplished by using the Branksome triangle or latterly, following withdrawal of rail services to Bournemouth West in September 1965, the turntable at Bournemouth shed.

This article originally appeared in the S&D Telegraph magazine for Autumn 2016.

The famous grounded coach body used for enginemens' "Mutual Improvement Classes" resists the tide of destruction engulfing the old Midland shed at Green Park. The turntable was cut up with indecent haste as early as October 1966. The shattered remains of the S&D's own wooden shed lie mangled and twisted on the left. Running lines still remain on the right as goods traffic continued to be handled in Midland Bridge Yard until June 1971 after which large numbers of condemned box vans were held here for the next year or so – the final working out of Bath being in April 1972.
PHOTO: JEFFERY GRAYER

LMS GAS STORAGE TANKS

TWO WAGONS OF THIS TYPE HAVE BEEN COMPLETED RECENTLY BY CHARLES ROBERTS & CO. AND A THIRD IS UNDER CONSTRUCTION.

The photograph (right) shows a gas storage tank wagon recently built by Charles Roberts & Co. Ltd., Horbury Junction, near Wakefield, to the specification and designs of the London Midland & Scottish Railway.

The principal dimensions of the vehicle are as follows:

Length of tanks outside	24ft
Diameter of tanks outside	7ft 6in
Length of headstocks	25ft
Length over buffers	28ft 5in
Height of buffers unloaded	3ft 5¼in
Wheelbase	15ft
Capacity of gasholders	958 cubic feet
Working pressure	160lb. per sq. in.
Cubic feet of gas at 11 atmospheres	10,538

The underframes are built of rolled steel channels, angle knees and plates. The buffers are of the Turton, Platts, self-contained type, made of forged weldless steel. The drawgear consists of two independent drawbars bearing on to rubber drawbar springs, the hooks being of the Gedges type fitted with screw couplings of the standard L.M.S. pattern. The bearing springs are of the carriage type, having seven plates 4 in. x $\frac{5}{8}$ in., centres of eyes 4 ft. 6 in. when straight, and fitted also with auxiliary helical springs 4¾ in. long when free, made of rectangular section steel. The vehicle is equipped with the automatic vacuum brake, and hand brake levers working in conjunction with the through brake shaft.

The wheels and axles are of the carriage type, having disc centres, the wheels being 3 ft. 7½ in. diameter on tread. The tanks are constructed of three-shell plates with two deep flanged ends, the shell plates being $\frac{11}{16}$ in. thick, and the ends ¾in thick. The circumferential seams are double riveted with rivets 1in diameter. All longitudinal seams specially formed with two rows of double riveting and an inside cover late to further strengthen the joint.

Each tank is fitted with three longitudinal rods 1¼in diameter, having 1½ in. diameter screw ends, and fitted in

the centre with turnbuckles and lock nuts, a manhole with cover plate and inside stiffener ring being arranged on the top of the tank at one end. Each tank is fitted on the outside with a hydrocarbon valve and pipe connection inside the tank of copper pipe $\frac{5}{8}$ in. diameter bore, bent to follow the radius of the tank to the centre line and at the lowermost point slightly turned up, for the emptying out of any deposit of hydrocarbon, etc. In addition, a similar type of valve is fitted on the outside of the tank, coupled to a range of high-pressure inch-iron piping, which is arranged to deliver the gas on either side of the underframe. These iron pipes are connected to two copper pipes on each side of the vehicle, one being connected to the pressure gauge, and the other to a gunmetal gas-filling valve carried on wrought-iron brackets secured to the solebars. These filling valve connections are provided with gunmetal caps with chains. The tank and piping was tested hydraulically to a pressure of 240lb per square inch after completion.

The tank mounting consists of 7 in. x 7 in. pitchpine cradles, running longitudinally with the tank with 12in by 10in end stops, 5 in. x 4 in. x ½ in. tee iron end stanchions, 9 in. x 3½ in. x $\frac{7}{16}$ in. end crossheads, diagonal tie rods 1¼ in. diameter, with 1½ in. diameter screw ends, and two holding down bands. The diagonal rods and holding down bands are fitted with lock nuts. The tare weight of these vehicles is 17 tons 8 cwt 2qrs., the weight of the tank alone being approximately 9 tons 10 cwt. Each tank is fitted with two lifting eyes to facilitate removal.

This article originally appeared in The Railway Engineer, August 1928 issue.

THE ONE AND ONLY 'BIG BERTHA'

A UNIQUE ENGINE DESIGNED FOR ONE PURPOSE

Midland Railway and LMS No. 2290/22290 (BR No. 58100 from January 1949) ceased to exist in September 1957, a long 65 years ago. As a one-off it might have been expected to fade from memory, destined to be forgotten by most, not least because it was rarely seen other than between Bromsgrove and Blackwell or between those points and Derby when works attention was required.

Its endearing fascination has to be its uniqueness, both in design and the job it performed, and performed it did – well – lasting in service for just under 37 years covering 838,856 miles in the process, a mean average of 22,671 annually; not bad for an engine whose daily task was shuttling the few miles between Bromsgrove and Blackwell before returning light to the starting point ready to repeat the process all over again – and again – and again.

Built at Derby to the design of (Sir) Henry Fowler, Midland Railway No. 2290 entered service in December 1919, an 0-10-0 tender engine specifically designed to assist trains up the fearsome 1:37.7 Lickey Incline of the Birmingham to Gloucester railway.

As to why it was built is perhaps slightly questionable for although No 2290 could certainly perform the work of two 0-6-0 tank engines undertaking the same task, the only real saving was in terms of one locomotive crew, the amount thus saved hardly likely to cover construction and maintenance costs of this single far more complicated machine.

For complicated she was, an 0-10-0 tender engine (were the centre driving wheels flangeless as on the later BR 9F 2-10-0 type?) and having four 'simple' cylinders. The two inside cylinders shared the same piston valve as the respective outside cylinder in the form of a cross-flow arrangement. This was necessary as there was insufficient room within the frames for four separate valves, one per cylinder. Technically it was a similar arrangement to a contemporary 0-10-0 compound operating on the Italian railways, drawings for the latter retained at Derby. As such there was some noticeable throttling of steam that occurred on No. 2290 but over the short distance she had to perform it was not considered too much of a disadvantage and she retained this arrangement throughout her working life.

Other than those already mentioned, facts and figures appertaining to the design are likely already known, but for completeness we might add the driving wheel diameter was 4 ft. 7½ in. and the boiler was pressed to 180 p.s.i.

As a one-off only ever performing in a very restricted location what might otherwise have been so easily lost was the day-to-day experience of working on the engine but fortunately this can to an extent be put right with abridged extracts from an article that appeared in the May 1956 issue of the now defunct 'Railway World' periodical. This was entitled *Return to Bromsgrove and the Lickey Incline* written, by T. P. Dalton.

Mr. Dalton starts, 'My first acquaintance with Bromsgrove was made 25 years ago, when I arrived with a somewhat heavy heart to start my boarding school days. At that time my locomotive observations had been confined to light engines operating on the ex-Cambrian Railways, and the thought of being within walking distance of the famous 'Lickey Banker' certainly helped to dispel some of the natural pangs of home sickness. No. 2290 as she was then numbered, was, alas, the cause of the cardinal mistake of being late for my first Sunday lunch at school. Needless to say, notwithstanding the appropriate punishment,

A British Railways view deliberately taken to show the efficiency of the front headlight fitted to No. 58100. BR(M) DY11750

As LMS No. 2290 at Bromsgrove awaiting its next turn. The number was altered to LMS No. 22290 in 1947 to avoid confusion with a new build Fairburn 2-6-4T. As to why the name 'Big Bertha' we cannot be certain. It certainly seems a strange copy of the name given to a huge German gun based in Belgium in WWI and able to reach the English shores. This is especially so with the origins of the name known to many railwaymen. Whatever, it was said with a degree of affection but was only ever an unofficial title, no name ever carried. Perhaps because of the continental connection she was also referred to as 'Big Emma' at Derby. TRANSPORT TREASURY

I returned to Bromsgrove South in the afternoon. The sight of the great engine coupled with my knowledge that she had a tractive effort of 43,315 lbs very soon counteracted the feeling of injustice within me.

'For the next five years I observed her from almost every possible vantage point and in all types of weather.

'Recently, in a far happier frame of mind, I returned to Bromsgrove courtesy of the London Midland Region, to make a series of footplate trips on 'Big Bertha' herself and some of her sisters, the 0-6-0T banking engines. No. 58100 is now something of a veteran and has spent her whole life at Bromsgrove. The fact she has received only slight alterations and run without a serious mechanical defect speaks volumes for her designer.

'Big Bertha' has acquired only a few alterations and modifications from her original form, one of these being the provision of a tender cab, essential one might say when running light back down the incline. Anyone who knows the Lickey must surely be familiar with the biting winds in the winter. My own limited footplate experience brought home to me the full advantage of this addition. In 1922 a large headlamp and turbo generator were fitted to assist in her picking out trains which required assistance up the incline after dark. Due to the effect of the General Strike and coal shortage, she was temporarily fitted for oil burning in 1926 and the next year the original Ramsbottom safety valve was replaced with the Ross 'pop' type. Then in 1938 the steam reverser was changed to hand-screw operating gear. I believe she has also had a new chimney.

'At Bromsgrove itself, in addition to No. 58100, there are seven 0-6-0T engines although not all would be in service every day. No. 58100 goes into Derby works annually and has a boiler washout and minor attention at Bromsgrove once every two weeks. The banking engines are normally to be found at Bromsgrove South, 640 yards from the main station and engine shed. Coal and watering facilities are provided in a dedicated engine siding at Bromsgrove South as well as a mess room for the banker crews.

'The fearsome incline is two miles four chains in length and begins immediately at the north end of Bromsgrove station. Fortunately it is dead straight throughout the climb. The incline is also a bit of bottleneck as there are four roads to the south and similarly beyond Blackwell, consequently trains cannot be allowed to fail on the incline for fear of creating serious delays to following traffic. In steam days most trains would require assistance, the varying weights of freight being the most difficult to assess. In fact the severity of the gradient meant that only the following maximum weights/consists were permitted to tackle the climb unaided.

Passenger – 90 tons including brake van/coach

Goods – 8 mineral wagons

12 general goods wagons

16 empty wagons

The above regardless of the type of locomotive.

'One 0-6-0T engine was stipulated to assist with loads up to 270 tons. From the latter weight up to 370 tons it was two 0-6-0Ts or No. 58100 and above 370 tons No. 58100 and one 0-6-0T. (There were some variations dependent upon the type of trains involved and meaning the weights could be reduced and greater assistance dictated according to the type of train.)

'The practical method of working the incline was for the driver of an approaching train to notify the signalman at Stoke Works Junction south of Bromsgrove of his requirements by means of a whistle code.

'1 pause 1' for one assisting engine

'2 pause 1' for two assisting engines

'3 pause 3' for three assisting engines

'If a train should fail on the incline it was the responsibility of the guard to walk back to the nearest signal box to request further assistance.

'Remarkably, during the whole of my school days I never witnessed a train fail on the incline: however this was to be reserved for my recent visit when I saw an 8F stalled with approximately 50 mineral wagons notwithstanding the presence of two 0-6-0T engines at the rear. Here undoubtedly was the difficulty in assessing the exact

In working guise. When banking No. 58100, as with the 0-6-0T type and their later steam replacements was always smokebox leading. This ensured the boiler water level remained at its maximum over the firebox crown on the gradient. TRANSPORT TREASURY

Rear detail at Bromsgrove, the various fire irons will be noted along with the 'dustbin-lid' type cover to the water filler. In the text Mr. Dalton comments about the liking for the engine by the Bromsgrove crews. To be honest this may have been as much about familiarity as anything else, especially when an interloper from another railway (the LNER Garratt) was thrust upon them. There may also have been a degree of pride involved knowing they had what was always a one-off. TRANSPORT TREASURY

weight, the majority of the wagons being empty. I do not know the full facts as I only witnessed this and the frustration on the faces of the crew on the lead locomotive, as at the time I was going the other way, down the incline, on No. 58100.' (Mr. Dalton does not elaborate on how the recalcitrant train was rescued.)

Mr. Dalton then makes an excuse that space does not permit a description of every trip he made but he does go on to describe three specific experiences. He starts, 'It was an extremely bitter November day with poor visibility when I made my way to Bromsgrove South and met Driver Randall and his young fireman, L. Wood, who were in charge of No. 58100. Driver Randall had spent 11 years at Bromsgrove. The thrills of riding on some of our main line express engines were certainly not to be repeated here, for instead the real interest lies in the working of trains up this massive incline. Of necessity it also involves a great deal of time standing idle in sidings waiting for the train requiring assistance. During these periods I was to get to know the men who carry out this less exciting but vitally important work. Bromsgrove men are also not confined just to banking turns as their duties can involving working south to Gloucester and as far north as Derby.

'Our first train requiring assistance was a freight of about 50 wagons with a 3F 0-6-0 at its head. The boiler pressure of No. 58100 was up to 180 psi and she was just blowing-off as we slowly moved out of the siding and gently came up to the rear of the freight train. There is of course no coupling-up and very soon, following the appropriate exchange of whistles, we were on the move and had got hold of the load with loud and crisp beats of the exhaust. There was a slight tendency to slip at first but No. 58100 soon settled down and I would estimate her speed to have been already about 15 m.p.h. as we passed through Bromsgrove station. The effect of the incline is most distinctly felt on passing the station and a gradual slowing down is noticed. Drivers tell me the steepest part is between the two over-bridges, a distance of perhaps 200 yards, and

Again at rest at Bromsgrove. The front handrails were likely to assist when cleaning the smokebox rather than having any connection so far as banking was concerned. Note the lack of any vacuum hose – implying the engine was fitted with just a steam brake (and a tender handbrake of course) plus the short framing ahead of the clamped smokebox door. A simple chain coupling was also fitted. TRANSPORT TREASURY

Work done, the gentle ride back. The actual ride is not mentioned by Mr. Dalton and this may be taken to mean it was not noteworthy. Certainly the engine would never have been likely to attain any great speed although due to the long coupled wheelbase may well have set up a 'shuffling' motion at times. TRANSPORT TREASURY

it was here that Driver Randall was working the engine about 45% cut off but with the regulator nowhere near fully open. Noticeable were the famed smoke-blackened trees and grass nearby. We were now approaching Blackwell and there was a sudden increase in speed as the train engine and wagons passed over the summit. Assistance does not stop here but continues through Blackwell station.

'No. 58100 now slowly drops away and moves into a siding, on this occasion to wait for two of the tank engines which had started to assist another train up the incline. Boiler pressure had also dropped by about 20 psi; we had taken about 17 minutes to make the ascent.

'Working back to Bromsgrove we pass a semaphore signal only applicable to banking engines which displays either 'C' or 'W'. 'C' stands for caution meaning there is another train in the section ahead – curiously the working is such that a banking engine may proceed down the gradient behind another train provided this is not a passenger working. 'W' indicates the section is clear and the banking engine(s) may proceed down the gradient up to their maximum permitted speed of 27 m.p.h. In addition any number of engines may proceed in this fashion down the incline at the same time and not coupled!' (Mr. Dalton does not mention it but this cannot have applied during poor visibility, perhaps also not at night.) Possibly the reason for this strange working being single line occupancy,

'The second experience was again on No. 58100 but this time assisting an 11 coach passenger train hauled by Black 5 No. 44746. Again we came up behind the train just prior to Bromsgrove station and started to push without any trace of a slip. Driver Randall was working the engine at about 30% cut off and we fairly roared through the station which

A final public viewing of No. 58100 at Deby on the occasion of the works open day. The engine had already been withdrawn, on 19th May 1956, reputedly due to a replacement boiler/firebox being required, both of which were of course non-standard and consequently with the cost not justified – note too the missing headlight already transferred to No. 92079. (The 3Fs were never fitted with a headlight as the driver was better placed to see the position of the rear of the train ahead due to the short length involved.) The board placed at the front described the principal dimensions along with the words, 'Formerly at Motive Power Depot Bromsgrove'. Even though it would never steam again the external condition is good, the engine having received lined BR black livery in April 1950 complete with the then BR crest. As with the Garratt types on BR, No. 58100 was never given a letter/number power classification. TRANSPORT TREASURY

needless to say gets its fair share of soot and smuts. As before very little firing was done on the incline. Fireman Wood had built up a good fire and when attention was needed the door remained open. Once more we lost about 20 psi on the climb but completed it in 9½ minutes. I calculated the engine had evaporated about 1,000 gallons of water.

'As a comparison the next trip was made on an 0-6-0T. The engine concerned was No. 47308 with Driver Miller and Fireman Eric Underhill. No. 47308 had also been inscribed 'Meccano Castle' on one of the side tanks. The train we were to assist was of five passenger coaches already double headed by a Class 5 and 2P. I am sure this train did not really require assistance but there was no brake van at the rear which meant a banker was automatically required.' (Would the same apply that the final passenger vehicle should be a guard's brake?) 'This service also called at Bromsgrove so the assistance started right at the foot of the bank. Naturally with such a light load we fairly romped up the bank in about 7½ minutes with no firing needed and boiler pressure remaining constant.

'During the wait at Blackwell and having to allow a

down passenger train to pass, I was able to get the views of the men on the various banking engines. Bromsgrove men all have a deep respect for No. 58100 and the little 0-6-0T types are similarly very popular. Reading between the lines the trial of the LNER Garratt No. 69999 was not popular, being very hungry and thirsty. The BR 9F tried, No. 92008, was recalled as a good engine for assisting passenger trains but was felt to have nothing in reserve compared with No. 58100.' (Ironically it was another 9F, No. 92079, that took up the mantle after the demise of 'Big Bertha' also taking over the formers front headlamp.)

'Mr. Dalton concluded by describing Bromsgrove shed as the cleanest he had ever visited, no doubt much due to the efforts of the shed master Mr. J. Wilkins.'

Just over 100 tons of scrap awaiting its destruction which would take place in September 1957. The boiler lagging and front numberplate have been removed whilst the cylinders were supposed to have been removed and saved, intended to show the unusual cross flow arrangement. In the event it is not believed these were retained. The reason No. 58100 remained a one-off was simply there was no need for further examples; as built it had proved a point that here was a suitable banking engine – also the largest Midland Railway engine ever built. It remains a great pity it was not saved for posterity. TRANSPORT TREASURY

BANGOR STATION

A STATION HEMMED IN BY TWO PROMINENT HILLS

Bangor station was originally part of the Chester & Holyhead Railway (CHR) which was authorised in 1844 and opened in stages between 1848 and 1850. The station opened on 1st May 1848, and for a period it was the western terminus of the line from Chester. The final section of the route to open was that between Bangor and Gaerwen (including Britannia Bridge), this occurred on 18th March 1850 making Bangor a through station.

The town is located to the north of the Snowdonian mountain range and is on the south side of the Menai Strait that separates Wales from the Island of Anglesey and lies between two prominent hills, the station is located in the space between them. To the east of the station the line passes into the 890 yard Bangor Tunnel, and to the west into the 648 yard Belmont Tunnel.

The station building was the work of railway architect Francis Thompson whose distinctive work graced the North Midland and Eastern Counties railways as well as the Chester & Holyhead, with access to the station via a driveway that connected the main entrance to Station Road, on the southern edge of the town centre. Although altered and extended since its construction in 1848 it carries Grade II listing. The station had goods facilities and an engine shed located to the south.

On 1st January 1859 the CHR became part of the London & North Western Railway (L&NWR), who later that year replaced the original engine shed with a larger structure.

Bangor became a busy trunk route, and with the opening of many branches in the area the facilities at the station quickly became inadequate. In 1884 the L&NWR enlarged the station. The down platform was enlarged and became an island through the addition of a platform face on its north side, with waiting rooms and an extensive canopy being provided. The alterations to the down platform required the demolition of the 1859 engine shed. The replacement was a six-road straight shed built just to the south of the station. A footbridge was also provided towards the western end of the platforms. On the up platform a bay had been created at the western end therefore giving a total of four platform faces. As seen in the photograph on the right, this gave room for two fast lines giving four tracks between the platforms.

To the south of the locomotive shed a large goods shed was provided with sidings and a traverser crane. A footbridge for staff linked the down island platform to the goods yard at the western end of the station. Signal boxes were provided at each end of the station. Bangor No. 1 was located on the south side at the eastern end, while Bangor No. 2 was situated on the north side of the line at the western end.

During the late Victorian and Edwardian period Bangor station was an extremely busy place and offered passengers a wide variety of services, linking the town with fast expresses directly to London, the Midlands and Holyhead. Other express trains provided connections to Liverpool, Manchester and Yorkshire. There were also frequent stopping services to Afon Wen, Amlwch, Bethesda, Chester, Holyhead and Llandudno.

On 1st January 1923 Bangor became part of the London Midland & Scottish Railway and between 1923 and 1927 they carried out major works at Bangor. The up platform was converted into an island when a northern face was created and two lines laid (one serving the platform and one as an avoiding line) along what had been part of the approach road. The original CHR building became isolated from the approach road on the up island platform. The solution was a new building that provided the main entrance and booking hall which was constructed north of the new line. Planning of the new entrance building was probably at an advanced stage during the final L&NWR years. Detached from the platforms, the design of the new brick building was in sympathy with the 1848 block, enhanced by a full-length awning of steel and glass, cantilevered from the building, the fascia member being decorated with roses. The entrance led to the broad

The revised layout at Bangor station showing the two through lines to good effect. TRANSPORT TREASURY

staircase and covered footbridge which connected the platforms. A new shortened approach road was created.

On the up platform new facilities were built in the Art Deco style adjacent to the original building. At platform level it was clad in tiles so that it blended in with the newer structures.

The new platform face on the up platform was numbered 1 and the original up platform was numbered 2. The original down platform became 3 and the southern face 4. A new bay platform, called the Bethesda bay, was created at the east end of the down platform for use by the local trains that served that destination.

Little changed at Bangor until the 1950s. On 3rd December 1951 the Bethesda service was withdrawn leaving little use for the Bethesda bay. In the 1960s many other local services were also axed. The run-down continued and the locomotive depot closed on 14th June 1965.

On 8th December 1968 Bangor No. 1 signal box was closed and demolished shortly after, with Bangor No. 2 box taking control of all lines, this was renamed simply as Bangor.

The northernmost platform face (platform 1 created between 1923 and 1927) had been taken out of use by 1980, and the track was lifted. This effectively returned the situation to how it had been. However the 1920s booking office was retained as the main entrance, and the trackbed adjacent to platform 1 was converted to a car park.

By 2013 Bangor station had only two platform faces having reverted almost to how it was in 1848. The locomotive depot building and goods shed were still standing but the area adjacent to them was developed as a station car park. A war memorial was relocated to the station (near the corner of the main building, Platform 1) in 2019. It commemorates former members of the Bangor Railway Institute Boys' Club who died in the First World War The station continues to run services to Birmingham, Cardiff, Crewe, London, Manchester and Holyhead.

VIROL
VIROL

A superb panorama of Bangor station on 27th July 1955 with five steam engines and a DMU in view. The only loco identifiable is the Fowler Class 3 2-6-2T No. 40003 at the bottom of the picture with a couple of Black Fives and a Fairburn tank. No. 40003 was one of a class of seventy engines designed by Sir Henry Fowler and introduced between 1930 and 1932, all being built at Derby. In 1948 No. 40003 was based at Rugby but had moved to Bangor in November 1955; allocated to its final shed at Widnes in 1960, it was withdrawn from there in February 1961 and scrapped at Ince Wagon Works, Wigan.

WEST END OF EDINBURGH

BY IAN LAMB

The London, Midland and Scottish Railway (LMS) was very much the dominant company in and around Glasgow before nationalisation in 1948, mainly due to the earlier sprawling efforts of the magnificent Caledonian Railway which it took over in 1923.

This was not the case where Edinburgh was concerned. Scotland's capital city had its railway tentacles well and truly all over the place, and they were very much the olive brown of the North British Railway, whose headquarters in London Road overlooked the ever inspiring setting of its wonderful Waverley station.

Not to be outdone, plus a determination to 'cash in' on Edinburgh's commercial potential, the Caledonian Railway laid an almost dead-straight line from the western outer suburbs of the city through and over the NBR network to the end of Edinburgh's main thoroughfare – Princes Street – and named its station accordingly.

ABOVE: A view of Princes Street station from a departing train which shows the curvature of the train shed that took passengers into the heart of Edinburgh.
PHOTO: NORRIS FORREST

Taking the NBR challenge in its stride, the Caledonian Railway went one stage further and built the amazing Gothic Caledonian Hotel above and astride the narrow arched entrance and curved fan of seven terminal platforms. Undoubtedly it was a marvel of Victorian engineering, and commonly known as the 'Caley' station. Indeed, where physical attraction is concerned, many would say that the architectural beauty of the red sandstone Princes Street station far surpassed the 'Waverley' complex lying in the bowels of the former 'Nor' Loch' at the other end of the street.

Temporary terminals were opened in 1848 and 1870, but it was 1890 before the main structure was truly open for business whilst the surrounding complex was greatly expanded and redeveloped. When completed, it was covered by an 850 feet glass canopy, overlooking vehicle parking and a large circulating area.

Traditionally, Princes Street station was chosen by visiting monarchs, statesmen and celebrities as the entry point to the Scottish Capital city because of its ground-level platforms, very much favoured over the steep inclines of the 'Waverley's' long ramps. After Queen Victoria, all of the 20th century British monarchs began their state visit processions from this west end terminus.

Whether we like it or not, our lives are increasingly being dominated by sport. For instance, if it's not football its tennis; if it's not athletics it's cricket, even though we may have no interest in any of them whatsoever. But, wait a moment. There was an exception where rugby was concerned from a train watcher's point of view because in the steam days of the 1950s, rugby involved many special trains for the 'fans' at Murrayfield.

Secondary school education in Edinburgh's Gorgie district introduced me to another important aspect of the Edinburgh railway scene, namely that of the former LMS which I'd certainly known about through almost annual trips down the 'West Coast' to relatives in Coventry via Birmingham New Street Station. I have always admired

29TH FEBRUARY 1964
Princes Street signal box. This massive signal box housed no less than 156 levers which controlled train movements in and around the terminus.
PHOTO:
NORRIS FORREST

12TH FEBRUARY 1965
The view looking out of Edinburgh Princes Street station.
PHOTO:
SANDY MURDOCH

21ST OCTOBER 1963 – On a wet evening, 'Black 5' No. 45476 waits at Princes Street station platform 4 with the 5.18pm for Glasgow Central whilst sister engine No. 45214 at the adjacent platform was at the head of the 5.32pm to Stirling. PHOTO: W.A.C. SMITH

19TH APRIL 1965 – Legendary Caledonian Railway 4-2-2 locomotive No. 123 with its appropriate coaches (preserved by the SRPS at Bo'ness) arrives at Princes Street Station with the Easter Weekend 'Scottish Rambler' railtour from Glasgow Central. This classic engine was built by Neilson & Company of Glasgow in 1886 and went on to show at the Edinburgh International Festival of the same year, whilst during the 'Race to the North' (the competition between the West and East Coast routes) in 1888 she made a record run from Carlisle at an average speed of about 60 m.p.h. No. 123 now resides in the Glasgow Museum of Transport. PHOTO: W.A.C. SMITH

29TH FEBRUARY 1964 – A typical 'dreich' Scottish weather afternoon hangs over Princes Street Station as Stanier 4-6-0 No. 44978 accelerates up the hill with the 4.22pm train for Perth. PHOTO: W. A.C. SMITH

Stanier's 'Jubilees'. My love for these 4-6-0s mainly came about by a daily sojourn after school to Dalry Road sheds to see what had brought in 'the Birmingham'. One of the most regular engines was 45620 *North Borneo* of Nottingham shed (16A), so that was the engine I modelled.

The beauty of the LMS compared to (my preferred) LNER was that the former's locomotives tended to be 'common users', so it was not strange to see 'Jubilees' in Edinburgh which had originated throughout the LMS section. This was particularly so with Rugby Internationals (especially Welsh) when 'spotting' at Dalry Road was simply paradise!

Despite the cramped site of the motive power depot (perfect modelling prototype), the adjacent station island platform was wide enough to cater for the local lines to Barnton and North Leith, and especially to Stirling where the former LMS had running rights over the NBR main line from Haymarket West, through Saughton Junction to Polmont.

Alas – as a railway station – the Princes Street station edifice only gave around a century of service when it closed on 9th September 1965. The rails were lifted, and the track base eventually became Edinburgh's much needed West Approach express roadway route in and out of the city. The hotel still exists though under a different name, so any reference to Caledonian or LMS now rests solely in the history books! This grand station was demolished within five years from 1970.

1957 – (17A) Derby's 'Jubilee' Class 6P 4-6-0 No. 45663 JERVIS is coaled at Dalry Road MPD. Having earlier arrived at Edinburgh Princes Street with the 11.25am train from Birmingham. PHOTO: G. M. STADDON/N. E. STEAD COLLECTION

BELOW: A map of Dalry Road station and engine sheds dating from 1932.

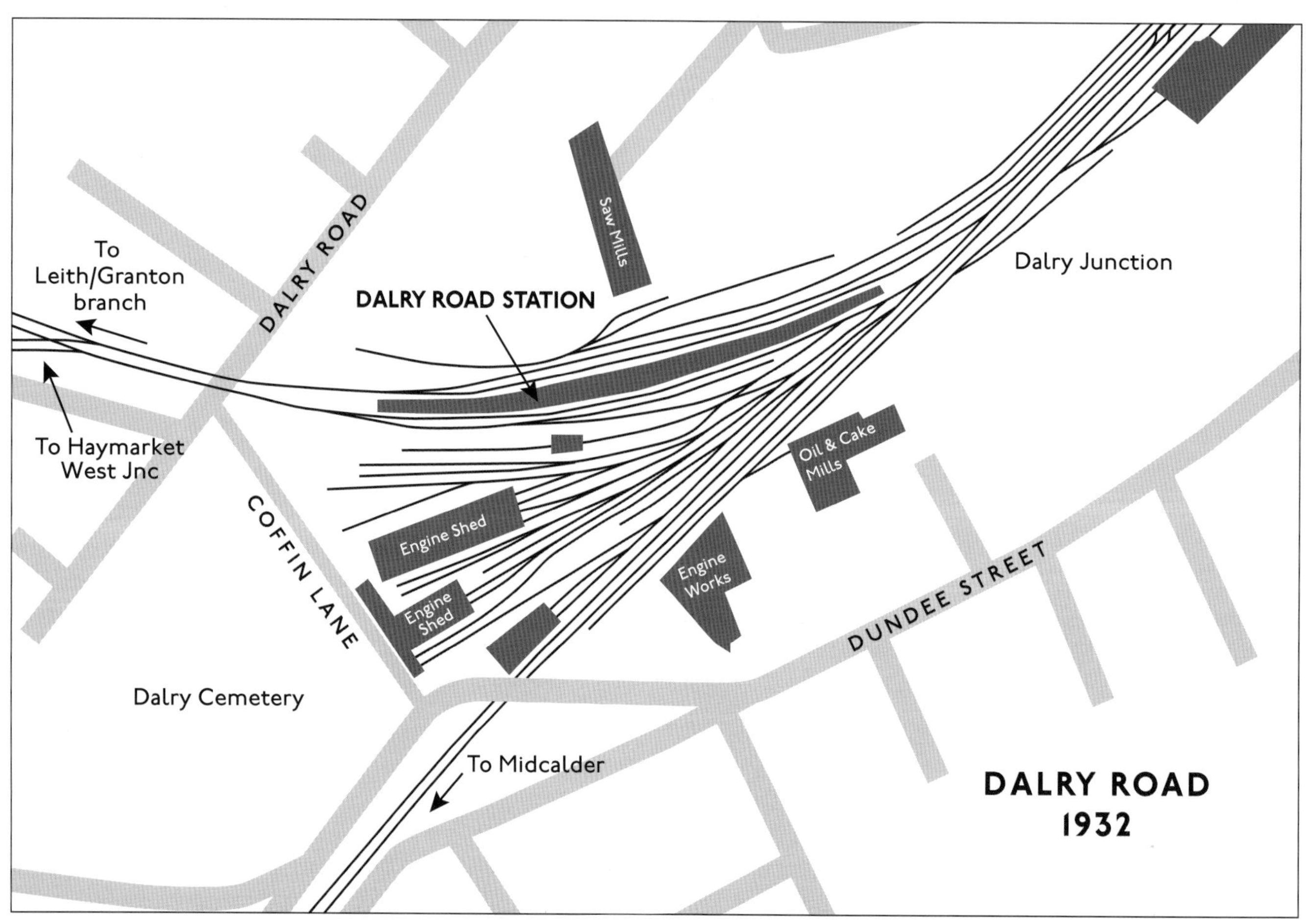

2ND FEBRUARY 1957 – This was a day of exceptional 'special' traffic. Pristine Class B1 4-6-0 No. 61117 of Parkhead (65C) arrives at Gorgie East station with the 11.51am football special from Glasgow for a match at Tynecastle. The rugby match at Murrayfield generated no less than nine specials at Gorgie East alone. PHOTO: W.A.C. SMITH

19TH MARCH 1960 – The running of dining car specials from Glasgow Central direct to Murrayfield Station on the occasion of rugby internationals dated back to Caledonian Railway days. These trains, loading to ten or more coaches and double-headed, used the line from Slateford to Coltbridge Junction, and on this occasion were hauled by BR Standard Class 5MT 4-6-0s Nos. 73060 and 73076 with the 12.40pm special from Glasgow. PHOTO: W.A.C. SMITH

29TH FEBRUARY 1964 – Dalry Road Motive Power Depot was situated three-quarters of a mile from Princes Street station where 'Black 5' No. 44994 passed through on the main line out of town with the 12.06 pm local train for Kingsknowe. Extravagant motive power for a three-coach train on a three mile journey! In addition to sister engines 45360 and 44953 in the middle foreground, there were 15 steam locomotives and three diesels in and around the shed on that date. PHOTO: W.A.C. SMITH

29TH FEBRUARY 1964 – 'Black Five' 4-6-0 No. 44952 crosses Slateford Viaduct with the 2.14pm train from Carstairs to Edinburgh which included through coaches from Manchester Victoria. PHOTO: W.A.C. SMITH

29TH FEBRUARY 1964 – Merchiston Station, 1¼ miles from Princes Street was opened for service in 1882. Stanier 'Black 5' 4-6-0 No. 44700 nears the city centre with the 3.05pm from Carstairs. PHOTO: W.A.C. SMITH

29TH FEBRUARY 1964 – Dalry Road (64C) Fairburn Class 4 2-6-4 tank locomotive No. 42273 arrives at the outer suburban station of Kingsknowe (originally named Slateford) with the 12.57pm local service from Princes Street. This station was closed four months later, but reopened in 1971. PHOTO: W.A.C. SMITH

6TH FEBRUARY 1965 – The Caledonian Railway completed a goods line from Slateford to Haymarket in 1853. With the closure of Princes Street station, this northern spur was revamped in September 1964 as the Duff Street connection. This enabled trains from Glasgow and Carstairs to reach the 'Waverley' without having to circumvent the 'Sub' line around the city requiring entrance to the station from the East. (9B) Stockport Edgeley 'Stanier' 5MT 4-6-0 No. 44867, fresh from overhaul at Cowlairs Works, heads the 1.28pm football special to Falkirk Grahamston. This junction is now known as Haymarket East and the spur was fully overhead electrified along with the Carstairs line. PHOTO: W.A.C. SMITH

THE PLATFORM END

In future issues our aim is to bring you many differing articles about the LMS, its constituent companies and the London Midland Region of British Railways. We hope to have gone some way to achieving this in Issue 1.

Midland Times welcomes constructive comment from readers either by way of additional information on subjects already published or suggestions for new topics that you would like to see addressed. The size and diversity of the LMS, due to it being comprised of many different companies each with their differing ways of operating, shows the complexity of the subject and we will endeavour to be as accurate as possible but would appreciate any comments to the contrary.

We want to use this final page as your platform for comment and discussion so please feel free to send your comments to: **midlandtimes1884@gmail.com** or write to **Midland Times, Transport Treasury Publishing Ltd., 16 Highworth Close, High Wycombe HP13 7PJ.**

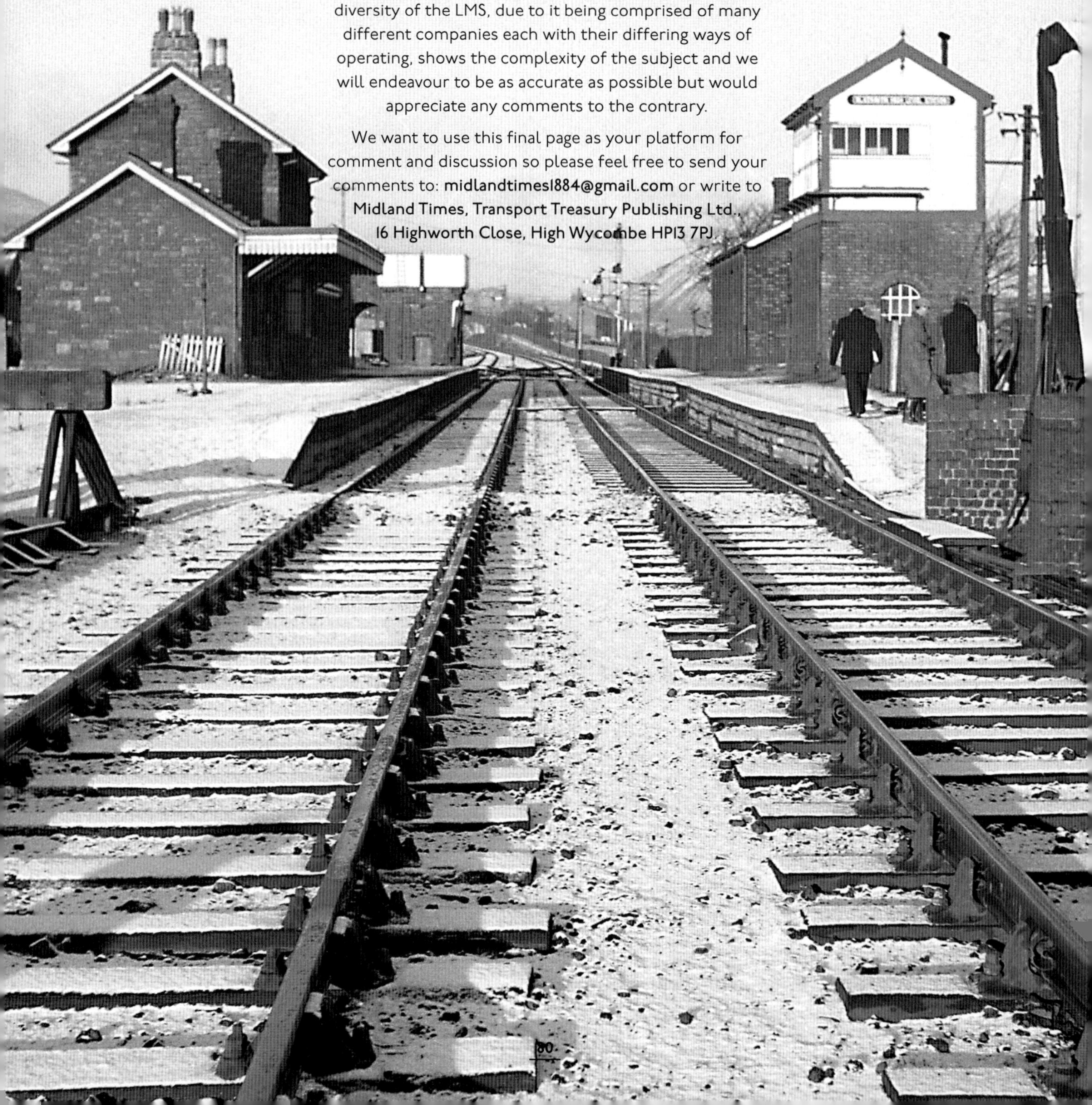